Software Testing Basics

Software Testing Fundamentals for All
Dedicated Testers

PAUL FELTEN

DEDICATION

To my beautiful wife and best friend, who never stops supporting me.

Contents

Introduction

I was lucky enough to start my software testing career with an apprenticeship under a well-known medical device expert and consultant. He took me under his wings throughout his regulatory consulting. I was able to see first-hand what companies were doing in the name of compliance with medical device standards and regulations. Each company has their own issues. I've seen very large companies that have cumbersome processes that must be followed without any room for error even though the process may not be applicable or valid and nobody was wise enough, or had the courage to challenge. I've also seen small companies that hired an inexperienced software tester (or forced that position onto an IT staffer) and is using common test and regulatory terminology incorrectly. When each person was asked why they adhere to a terrible or burdensome practice, almost all of them simply state that they were taught by their predecessor to do it, or that they inherited it and didn't think to change it. The few others simply didn't have a software test foundation to lean on. After directly being involved with software testing and verification for a half dozen clients, I completed my apprenticeship and continued with the same job – just with a little less oversight! It continues to be a wild ride through test automation and tailoring software test help to each client's needs, but it is completely worth it. I don't have a formal software testing background but my diverse background and experience gives me a unique perspective when accomplishing my work. Through this perspective, I see a training failure within the software testing community. Employees and consultants alike are misusing commonly used and defined vocabulary. Software requirements are not appropriate or clear. Test cases lack needed information and take exponentially more time to execute than they should. I could go on and on, but my point is clear. Entry-level software testers, senior software testers, test managers, consultants, and indirect software testing roles including product managers all need help to understand software testing basics. Once the basics are understood, then processes can be refined and updated, test templates can be streamlined, and requirements can be better clarified. Test efficiency can be maximized and more advanced knowledge can be pursued.

There is often a lack of material for new software testers. When I started my apprenticeship, I had no beginner knowledge as I had a bachelor's degree in history as well as US Army experience. My first task was to attend an AAMI course for the FDA's Global Principals of Software Validation. I don't remember any of it because the vocabulary and concepts were foreign to me, but it did introduce me to commonly used words and industry vernacular. I wish I could have found a software testing primer, to give me a foundation so I could have applied more of the training. Software testing foundational certification syllabi exist and advanced testing manuals but nothing in a shorter book form, to help the working professional.

My goal is to give current and future software testers a solid foundation and basic understanding of software testing best practices and processes so software testing can become efficient and effective.

Let's be very honest. Software testing is a very time demanding profession which leaves very little, if any, free time for professional development. As a software tester, I understand this to be very true. Very few software testers want to spend all day working just to come home and read a software testing book before falling asleep on page 5. This is the exact reason why this book is brief and precise. It should give you a quick, clear, and concise software testing foundation so you can remember it better. Please feel free to use this book as a desk reference as well. There are items within that are great to take to review meetings, to confirm definitions, and to implement into your own testing practices.

Lastly, this book derives its practices and standards from the medical device industry, which recognizes several elements that inherently create a better medical device. If these elements are applied to non-medical device software, quality will increase and better software will be delivered to production. Keep in mind that all elements may not be applicable to your product. I will never advocate for a one-size-fits-all approach, rather by reading this book you will see general standards for good software testing and a lean risk-based approach. Neither the author, nor publisher are making any claims that following this book will provide compliance for any company, product, or person. Regulated software testing does not provide a uniform approach, and as such, the author and publish are not liable for damages or misuse from improperly applying any words within this book.

To aid in understanding and to provide definitions of common testing terms the ASTQB® glossary is included with some FDA definitions for referencing.

1 FAQ

What is Software Testing?

The most simplistic definition of software testing is the process of evaluating and reporting the quality of software.

The definition seems simple enough, but the reality is not. Software testing consists of many activities and there are never enough time or resources to test software and complete test activities adequately. Similar issues can occur when no dedicated software testers exist, or the number of software developers are greater than the number of dedicated software testers which can create a bottleneck and place a large burden on each tester to be able to do more with less time.

Why Test Software?

Humans err, and imperfect people cannot create perfect software. If you doubt this concept, then try to develop a program without errors. This is why software testing is necessary. There are four specific reasons to test software; to give early feedback to software developers, to provide quality information to all stakeholders, to gain product confidence (or lack thereof), and to adhere to regulatory conformance.

Providing early feedback allows developers to fix

and improve software earlier, which can result in higher-quality software. This is primarily finding and reporting bugs. Providing early feedback ultimately depends on software developers giving software testers early and often access to updated software.

Quantitative metrics is test information provided to stakeholders so educated decisions can be made and justified. Stakeholders are people in positions that make product decisions, and have a responsibility to make decisions based on test feedback. Reported metrics are usually communicated through the number of bugs found, bugs remaining, and how many tests pass or fail.

Product confidence is directly related to product quality. When many bugs are found and the product is close to the release date, product confidence is low. Conversely, when most bugs are fixed and not many bugs are being found, then product confidence will be very high. Ask yourself, "based on the latest testing that was conducted on the software… can I trust the software to do what it is supposed to?" If you answer no, then you do not likely have much product confidence. Keep in mind that product confidence should change throughout the product development – hopefully for better.

Regulatory conformance is inclusive of any applicable standards a company must comply with. This could be company policies and procedures, international standards, or even FDA regulations. For most regulatory conformance, a number of deliverables are required to be followed, updated, and documented. Software testing usually has a large responsibility in achieving compliance.

What does Software Testing Look Like?

Software testing activities occur throughout all stages of the product lifecycle. On any given day, a software tester will usually test software, report bugs, and create, update, and review documentation – providing feedback when necessary. Usually at least one software tester will always be in attendance for meetings that directly or indirectly involve software testing such as reviewing requirements.

An entry-level software tester will most likely complete product orientation, learn company procedures for software testing activities, and run some menial and very long tests that nobody else wants to. Other new software testers will get a fresh look at the software in order to find new bugs that were overlooked.

A test manager will likely delegate test activities to subordinates and be focused on documentation such as planning and updating stakeholders with a test status. They also follow up on issues that prevent or inhibit test activities and attend meetings for reviewing requirements and discussing product schedules.

What Should be Tested?

Tests should not be created arbitrarily. Tests should focus on and be based on written requirements and risk. Requirements associated with a higher risk should be tested more extensively and with more rigor while low risk requirements can be tested with less rigor. Without requirements, testing can be aimless and doesn't accomplish much.

2 The Perfect Software Tester

One of the best aspects of software testing is the mentality of 'test to break'. Literally, the goal is to cause software to malfunction through various user-methods. For example, my favorite test is to type as many characters into an input field as possible and then try to submit that information. I have found some incredibly damaging bugs by doing this. You should never intentionally and maliciously break software, but should make sure that the software will work for the intended user, function as expected, and have consistent behavior throughout.

An analytical mentality is also important to know how to go about testing certain software functions. This is otherwise known as the engineering mindset – the ability to incrementally and sequentially work through, analyze, and create workflows and processes. This can be observed through creating test cases which tell the software tester what to do step-by-step.

In addition to the test mentality, there are many attributes that a tester must have, including but not limited to, attention to detail, trustworthiness, and excellent communication.

Attention to detail is one of the most important attributes a tester should have. Many times, a test will require reading through pages upon pages of text to verify that each letter and grammatical item is correct. This can be tedious but imperative to finding bugs and ensuring that the software is of high quality.

Trust is also imperative. Stakeholders rely on accurate test results to make important decisions and need to be able to trust that the tester provided the correct information. This also applies to reporting (or not reporting) bugs. Not reporting a bug due to laziness is dishonest and potentially withholds valuable information from stakeholders.

Excellent communication is mandatory. Much of the tester's time is spent communicating with managers, developers, and fellow testers. All communication needs to be clearly understood and written in a timely fashion.

A practical example of an ideal software tester is someone whom can learn quickly, tries to break the product by pretending to be the user, can visualize how to write a test incrementally, and can communicate clearly in a common language of the workplace.

3 Communication

Communication is a skill that directly affects software testing both verbally and in written form. Bad communication has the potential to waste valuable time on a project based on simple misunderstandings. Thankfully, there are many ways that testers can help overcome this communication barrier.

It's best when a software tester can communicate in the same language and without language barriers with the software developers and product manager. Native language skills are essential because it allows the tester to clearly communicate issues and timelines and also listen to software implementation and participate in important discussions when a tester's input is critical. It also diminishes the time needed for clarification. Instead of a one-hour meeting to discuss requirement changes, the same meeting can drag out to over 2 hours because of needing to repeat one's self over and over again to the group. This language barrier is also exasperated over the phone or via web conferences when static and poor equipment is used.

An extensive vocabulary is also necessary. A tester needs to know when to use what word, and why. For example, think about the word "message" and all the synonyms for such a word. "Notification", "text", "words", "phrase", "pop-up", and "sentence" are just a few. A good tester will think about the best wording for a test step before using it without wasting hours debating between the perfect wording.

An extensive vocabulary is also needed to rephrase for clarification. I was in a large meeting with a client who was discussing a critical bug that was just found. The software tester who found the bug used language that was common to the software test department but not common to the Regulatory Affairs attendee and product manager. After many attempts of saying the same thing, the test manager chimed in and restated the issue using more generic language and successfully conveyed the bug's criticality. Another example is a tester who writes a bug report but during a meeting has to explain that bug in a different way because the product manager doesn't understand what was written. This is not to say that the bug report was written poorly, rather, some people need alternative language to properly understand. It is vital that the tester be able to back themselves up with alternate wording.

A software tester's language must also be non-accusatory. This is one of hardest communication techniques to master. After all, as a software tester, I'm very good at pointing out what is wrong in software. It's what we do! But we need to be cautious to not offend and finger point, as this could degrade inter-departmental communication and hamper the project status and schedule. For example, "Johnny Programmer didn't program the software correctly. He misspelled three different buttons". This should never be said or written, no matter how true it may be. Johnny will not be happy and probably won't work with you in the future. A better option might be, "The text on three different buttons do not match what is written in the requirement". This second option is much better because it leaves out any names and simply states that the text is not what is expected. As awkward as it sounds, it's much better to speak about the software as a living being with fault, rather than blaming actual human beings.

Lastly, communication must factual and timely. Only facts and truth must be given to all stakeholders. Any lies can hamper product safety and confidence. Two of the worst lies that I have encountered as a software tester are software testers who 'pass' test steps without actually running the steps, and not reporting bugs that were found. Whatever reasons the software testers had were moot. They were found out and their reputation diminished greatly. Along with factual communication, timely communication is needed in order to keep on task and not block other people from working on a particular task. One client consistently took over one month to respond to any of my emails. To mitigate, I tried to only include one question per email in bolded font to no avail. Interestingly enough, all I needed was a written "yes" to complete testing, but I never received it (yes – I tried calling too!). I was embedded in another client's software test team who had four different written communication methods. Each was introduced as a new time-saving method, yet the time it took to stay updated with all four applications was greater than just having one method.

4 Test Independence

A lot of mythical software test knowledge involves off-base notions of test independence. Some think that a test executor cannot be the same as the test author. Others maintain that a programmer can also functionally test the software they programmed. Test independence is neither; it means that software developers should not functionally test their own software, which is not an excuse for software developers to avoid unit testing. It doesn't matter if the software tester that authored a test is the same tester that executes the test because ideally, that test should be reviewed, approved, and signed-off at a higher level – most likely by a peer or test manager. There is no conflict of interest, nor is there any subjective data that can be skewed by doing this. This is evident because the software tester's job is to find and report bugs. There is no conflict of interest in the testers role that would prevent them from reporting bugs.

There is however a conflict of interest when a developer formally and functionally tests the software they programmed. Everyone whom does their job wants to look good and be known for producing great results, but this conflict of interest is prevented by not having a developer functionally test their authored software. It's eliminating the potential problem of hiding or not reporting bugs found for the sake of looking good.

5 Software Requirements

Software requirements are the foundation of all testing and should be what tests are based on. They can also make or break a product and the product schedule. Companies with great requirements, more often release their software on-time with fewer bugs. Conversely, companies with poor requirements usually release their product months after initial timelines with lower confidence.

Software requirements are statements of what the software "shall" do. "Shall" is used because it's a definitive statement that clearly indicates what is expected, and what is not. "Will" or "must" are sometimes used, but "shall" is the most common and the best practice. Also recommended is to only use one of the three above words. "Shall", "will", and "must" should not be intermingled for consistency and predictability.

To complement consistent wording, all requirements should have consistent and positive language if possible. Consistent language will enable a clear interpretation of each requirement and enable requirement readers to predict what future requirements will look like. Positive language provides for clearer test verifications (expected results) to be defined from the text. Additionally, negative language should not be omitted from requirements, rather used sparsely and only when necessary.

Not only should every requirement use consistent and positive language, but every requirement should have a unique identifier or "tag" for easy identification and traceability. These are generally understood to be a necessity in order to confirm that all requirements are tested. It also helps to identify what tests will need to be updated when a requirement changes. Tags are also a form of verification and should never be changed unless it the same tag was found for a different requirement. Changing tags can result in hours of needless and duplicitous work. For example, one such client had three different requirements with the same tag number. The tags had to be updated as well as the tests that covered those requirements. The time it took to update those items was wasted and wouldn't have been needed if the requirement tags were not duplicated. If the software requirements are maintained in a word processing document, ensure that the requirement tags are not assigned by an automated numbering or lettering system within the word processing software. These automated features will change identification tags very frequently, which usually means that tests have to be constantly updated with the correct requirement tags. Take note of the below example below of a good requirement.

"[SRS-001] The software shall accept a username for registration, 6-24 characters in length."

The above requirement example contains all best practices previously mentioned. It contains a unique requirement tag ("[SRS-001]"), a "shall" statement, positive language, and identifies what the software does. A tester clearly knows what to test, what will pass, what will fail, and can begin to write a test for this requirement. When writing the test, the software tester should include test steps to verify a username with six characters and 24 characters is valid, while a username with five characters and 25 characters is invalid.

Alternatively, consider the poorly constructed requirement below:

"4.3.2 The software will not accept a username less than 6 characters and must not more than 24."

The above example contains many egregious errors. The very first error is the requirement tagging scheme - it appears that an automated numbering scheme is used. If a requirement is removed prior to "4.3.2", it's possible that this requirement will be changed to "4.3.1" which will require updating many references. Secondly, "will" and "must" are used inconsistently. "Shall" is preferred, but most importantly the requirement should be consistent and pick one of the two options. Another error is the negative language that is used. This negative language should be avoided when possible. The tester reviewing this requirement, preparing to write a test, will undoubtedly make sure that a username less than six characters, and greater than twenty-four characters is not allowed. But what *is* allowed? The requirement does not specifically state that a username with a certain length will be allowed, which makes the interpretation of the requirement difficult, and creates uncertainly.

After software requirements have been drafted, they should be reviewed. Requirements can be reviewed informally by testers when creating tests based on them. Frequently, when writing tests, software testers are able to find issues with requirements that need to be addressed and clarified before finalization and approval. Software requirements should also be formally reviewed and approved so individuals working with or testing the software know exactly how the software should be implemented at any given point.

Many times, software requirements will change throughout the product development, even last-minute. These changes should be encouraged even at the expense of test resources because the requirements will most likely be better worded and provide clarity. These requirements changes will frequently occur as a result of design review meetings in which applicable stakeholders meet to discuss the product design and any applicable updates. The updates can be to existing requirements or result in the creation of more requirements. This can include deferring a requirement to a later software version, retiring old requirements, or updating the text for clarity. Stakeholders in the collaborative review normally include one representative from the following departments, if possible: Test Team, Development Team, Quality Assurance, Regulatory Affairs, and the Product Manager.

The following software requirements review criteria should be considered when participating in an active reviewer role:

Unambiguous: It should be very clear what behavior is expected; what software behavior would pass and fail the requirement for a given test. Generic or unclear wording should not be used.

Specific: Accurate, not conflicting with other requirements. If specific measurements are used, a tolerance percentage or value should also be presented.

Testable: It should be clear how the requirement can be appropriately tested.

Traceable: The requirement should have a specific and unique identification tag to be referenced. A requirement's tag should remain constant throughout the software lifecycle to maintain traceability.

It's important to note the difference between system requirements software requirements. Software requirements define what the software does and system requirements define what the system does. They system is composed of all software elements. Let's imagine that the product you are software testing is a heart monitor. The software requirements would describe how the heart monitor software functions, like what is displayed on the screen at various times under various conditions. Now let's imagine that the heart monitor connects with a PC. Software monitors the heart in some hardware, and software transfers the data to an application on the PC. The system is the combination of all software. These requirements are higher in level and contain less detail whereas the software requirements are more specific. Remember that software requirements should (most of the time) come directly from system requirements.

Software requirement specifications (SRS) are documents that contain one or more requirements. Generally, there is an SRS that contains software requirements for all software, and then system requirement specification (SSRS) for all system requirements, but they can be combined if clearly labelled and defined. There are many ways to document these differences and not all of them are wrong.

The SRS and SSRS should be divided into similar subsections for clarity and to help reviewers and auditors easily find the information they are looking for. One such organization is to use software features as headings, for example:

- Login/Logout

- Search

- Feature 3

- Feature 4

6 Test Cases

A test case is a set of numbered and written instructions with expected results. It provides labeled fields for the recording of specific and needed test metrics such as the date, time, tester, software version number, equipment, serial numbers, and comments for both test deviations and any software bugs found. Its recommended that all test records include the test result, anomalies found, and tester identification which is the minimum information that should be present, but the previously listed information is the best practice for test identification.

Several layouts of test cases exist, some are good and some are bad. Specific examples are not provided, but important concepts are. When using a template, or creating your own template, consider the following:

Test Efficiency: Ensure a tester can easily flow through the steps without jumping back and forth between sections and without having to receive lengthy training. An efficient test will trim down test execution time which means that more tests can be run in less time. I was on-site with a client and they showed me an example test case from a different consultant whom specialized in hardware manufacturing. Each test had six sections and each section had different and conflicting instructions. With the many test case templates I have seen and used, this was the worst, and would have taken ten times longer to execute than a better template.

Verifications: Ensure that there is only one verification per step. Multiple verification statements can cause confusion, especially when filling out test results. The verification should be explicitly stated both in the actual step and in the expected results. Although this creates a longer test document, it provides a high level of clarity for the tester.

Traceability: Ensure each step or test case is linked to at least one requirement tag. As indicated, there are two main ways of accomplishing traceability within test cases. The first option is to list requirements being tested within the test case header. This can make for cleaner test cases with less information displayed for each step, however, the downside is that when a requirement is updated, the test in its entirety will need to be reviewed before updating the steps to match the updated requirement. The second option is to list each requirement being tested next to each applicable step. This can make each test step feel more cluttered, but saves time when updating test steps for changed requirements. It also aides in detecting any test coverage gaps.

Expected Results: Each step should explicitly state what the expected result is, and the expected result should mirror the verification statement in each step. For example, if the end of step one states, "verify a warning is displayed", then the expected result column for step one should state, "A warning is displayed". Although slightly redundant, this reduces confusion when executing tests.

Pass/Fail Results: Each step should be able to be clearly marked and distinguished between "Pass" and "Fail". Consider the spacing and size of the pass/fail results. If they are too close together, it may become unclear which is selected. One good approach is to use pass/fail check boxes.

Accountability: Ensure the tester has a labelled space to legibly print and sign their name at the time of testing. Also important is to allow room for a test reviewer to sign and print their name with the overall passing or failing test result. Keep in mind that just because a step fails does not mean that the entire test case fails. Possible test step failures which could still have a passing test case can be from a failing setup step or even a typo in a test step.

Updates/Revisions: Updates made to the test case after initial approval should be reviewed and approved again so a clear test case revision needs to be present. It's important to know whether you are executing "rev A." or a subsequent revision. Also, keep specific revisions/versions of software being tested out of the written test case. These versions should be listed in the test plan and test results, not the test itself. This will prevent future test case maintenance, or "churn", which will certainly require reviewing, approving, and signing off on the document many times over.

There are two main formats for recording test results; the merged format, and the un-merged format. The merged format includes sections within each test step for the passing or failing test result, as well as any comments or deviations taken during the test. This format has the benefit of only needing to maintain a single document. When the test is executed, the document is printed, then filled out. The disadvantage is that the document will have to be reviewed, approved, and signed-off multiple times. On standard computer paper, the test steps can be massively cluttered and hard to read. Also, if a test needs to be re-executed, then the entire test document needs to be re-printed and re-filled out which can use considerable printing resources. Conversely, the un-merged format consists of two documents. The first document is the written test itself which can be reviewed, approved, and signed-off independently from the test results. These test steps can be viewed on a computer or printed. The second document would then be printed and would contain many blank sections for all test information to record. These results would then be filled out during test execution, then reviewed, approved, and signed-off by a test approver.

7 Test Development

The test development process can change between products and companies, but all should include tasks within the process for ensuring test quality and accountability.

The following is an example of a test development process.

- Requirement(s) Received
- Test Creation & Traceability
- Test Dry Run & Debugging
- Test Review, Approval, and Sign-off
(if test is not approved, go back one step)
- Test Execution (Formal)
- Test Maintenance
- Test Retirement

Requirements Received

A test's foundation is the requirement(s) it is based on. In order to create a meaningful test, some sort of requirement must be documented, drafted, or verbally agreed upon. In Agile development and testing, user stories are agreed upon and written as requirements in a small round-table meeting. For waterfall type development, a design review meeting is carried out and all participants agree on a set of requirements. A tester must not wait until a requirement is formalized to start creating tests because requirements are likely to change throughout the product lifecycle and time is already limited so tests must be drafted as soon as possible.

Test Creation

After a requirement is received, even in the drafted form, a test case can be drafted. Usually, good test cases will have a test approach written first, followed by the test steps.

The test approach should provide anyone the ability to quickly read and determine what the test is accomplishing by including a verification statement for every requirement being tested. An example test approach is the following:

"This test verifies that the login button is enabled only when a correctly formatted username is entered. It also verifies…"

After the test approach has been drafted, the test steps are then drafted. Specificity and generality must be considered when writing test steps. Generality will cause the software tester to not write super-specific steps such that more elements and paths can be tested providing for better software quality. If the step is incredibly vague, then the tester should be directed within the step to record specific actions taken. Specificity in a test step enables non-testers the ability to execute tests. This can provide for flexible resources, or non-software testers, to execute test cases and speed up execution time. Since I generally hand-off my work to a client, I write very specific test steps with an occasional general test step to exercise the software better.

Traceability

Traceability may occur prior to drafting the test steps, but must be included in the test development process. Each requirement being covered in a test case must be documented. Usually this documentation is included in the test summary or a separate traceability document, although some include this information at the top of the test document. Often, the requirement being covered is documented in each applicable test step or in the test approach.

Test Dry Run

After the test has been drafted, it is more than appropriate to dry run the test case to make sure that the steps flow from one to another, to correct spelling errors, and to report any bugs that are found. This means non-formally executing the test case, running through each step in-sequence as if formally executing.

Test Debugging

During the dry run, the test author should fix any

test issues that arise. If there is a test issue and the approach is incorrect, the tester must re-author the approach and update the steps to match. Once the debugging is complete and the test is dry ran to the satisfaction of the author, the test proceeds to the next step.

Test Review

A test review is conducted and can be accomplished any number of ways. A peer, subordinate, manager, test lead, or even developer can review the test. Usually the review ensures that the approach and steps contain enough coverage for the requirement(s) being tested and that there are no test errors when executing the test. If the test passes review, then the test goes through the approval process. If updates are needed to the test, then the test is returned to the test author with specific notes for change. Once the changes are made, the test is submitted for review again.

Test Approval

If the test review is successful, the test is usually approved by a set of established standards. This is accomplished in a manual process by the printed name, date, and signature of the approving authority. There is usually one area in each test document used for a signature, date of the approving authority.

Test Maintenance

Requirements usually change, even after approved. This can be for many reasons, but not limited to:

- Inability of developers to implement
- Not enough time for full implementation
- Conforming the requirement to the current implementation
- Requirement does not correctly convey intended use

When a requirement changes, the test must also change to ensure that the test basis is being covered correctly. This means that the test goes back to the "Test Creation" part of the process and proceeds through the entire process once again.

Test Retirement

When a product releases subsequent versions and older requirements are no longer desired, the requirements are retired and are preserved, but marked appropriately to ensure that they should not be tested in the new version or product.

When a requirement is retired, tests must be executed to prove that the feature is no longer in the version or product. Once tested and proven, the tests or test steps may be removed or retired accordingly.

8 Test Types

Several types of testing exist and are defined below. New test types seem to be invented each year, but frequently are test types that previously existed, just renamed. The following testing types may slightly differ in other sources, but the intent of this section is to make the terms as clear and understandable as possible, while describing the correct use.

Acceptance: Testing that the software meets predefined and documented user needs.

Ad-hoc: Testing without written instructions.

Beta: Testing a product before it is released. This is usually completed by non-company affiliated personnel.

Black Box: Testing without knowing how the software is specifically programmed.

Boundary: Testing the minimum and maximum limits. For example, testing the maximum and minimum number of characters accepted in an input field.

End-to-End: Testing a whole product, usually as a user at the system level. Normally this is a test that mimics a user's entire workflow through the product.

Error: Testing how software responds to an error,

and/or attempting to defeat an error.

Experience: Testing software based on the tester's experience with the software, or similar software. This is usually related to ad-hoc testing.

Functional: Testing how the software functions. This is usually through the testing of the graphical user interface and content within. This occurs at a high level.

Integration: Testing to ensure that multiple software components work together as expected. For example, verifying that clicking a login button uses the correct username and password to log a user in.

Performance: Testing that software is usable and reacts within predefined speeds when a certain number of users or actions are being performed.

Platform: Testing that software is usable on supported mediums. For example, ensuring a website works on different browsers and operating systems.

Regression: Retesting to make sure a bug was fixed correctly, or to make sure no new bugs have been introduced into the software since last tested.

Stress: Testing the software at its maximum limits. For example, if a maximum of 1000 users should be supported by a website; testing to verify 1000+ users can use the website at one time.

System: Testing the product as a whole. This usually involves testing on customer devices using release candidate software. Many times, this will be end-to-end testing.

Unit: Testing the software code, from within the code. This is a developer's responsibility.

White Box: Testing, knowing how the software is programmed. For example, knowing where a database is stored for some software and manipulating the data in the database.

One or more test types should be used for any given test. The more test types used for testing a product, the better quality your tests will have, the more bugs you are likely to find, and the better test coverage you will have.

9 Test Levels

There are two main test levels: high, and low.

If a test is high-level, it focuses on the user perspective. It usually doesn't test specific software components and is more closely associated with system, black-box, functional, and end-to-end tests.

A low-level test has intimate knowledge of the software code and interacts with or views specific components to help test more rigorously.

System testing is generally the highest level of software testing while unit testing is the lowest.

High-level Test.
Step 1: Enter a valid username into the username field.
Step 2: Enter a valid password into the password field.
Step 3: Click the login button.
Step 4: Verify the user is logged in.

Low-level Test.
Step 1: Enter an invalid username into the username field.
Step 2: Enter an invalid password into the password field.
Step 3: Disable the user database.
Step 4: Click the login button.

Step 5: Verify the following error message is displayed, "…".

10 Objective Evidence

One of the biggest misinterpretations of software testing is "objective evidence". To many, this term exudes "screenshots". This is their final answer – hands down, and nothing can change that. They feel that screenshots, or pictures, must be taken every single step to provide objective evidence. Unfortunately, this couldn't be further from the truth. Objective evidence is simply showing that software requirements have been implemented and fulfilled which is usually displayed in a "PASS" or "FAIL" result.

Consider if screenshots are actually 'objective'. They are an image of the test step that was just executed by the tester. If the tester has enough lenience within the test step to be subjective, I would argue that not only is the test is not appropriate but the requirements are not appropriate either. So, if a screenshot was taken of what the subjective software tester executed of the subjective test step with subjective requirements, then is the screenshot really objective? No, it's not. It's subjective evidence.

When reviewing tests I receive from clients, sometimes I see each step state "take a screenshot and attach it to the test results". When manually executing test steps, taking a screenshot for each stop is not only unnecessary, but wastes a lot of valuable time. Think about how much time it realistically takes to take a screenshot, label it, print, then attach to the test results. It could add two minutes or more to a test step that should only take two seconds.

Knowing now that screenshots are not a replacement for or indicative of objective evidence, consider what screenshots (or pictures) can be used for. They can be used to provide accountability of the tester to the test reviewer. They can also help the reviewer to ensure that the step with a screenshot or picture did indeed line up with the pass/fail criteria.

Screenshots or pictures should not be eliminated from all projects, however, I recommended that screenshots or pictures only be used for individual test steps that are directly traced to software risks or risk control measures that if failed, could kill or significantly injure a patient.

11 Test Execution

There are two types of manual execution; dry running and formal execution.

Dry Run

Dry running is non-formally executing a written test. There are many reasons to dry run a test including reviewing someone else's test, making sure that a piece of software is not broken, and making sure there are no surprises during formal execution.

Documentation of the test results should exist, but may not need to go "on the record". For example, dry running a test for review would look like one employee writing a test case, then giving it to a peer. That peer would go through each step and pretend to formally run the test, while actually performing and possibly even recording the results of each action. Through dry running, the peer would give the test results back to the author with a review summary, indicating if it was a good or bad test, what could be improved, what steps were not clear, and what failed. The author would then make the changes and likely go to another (or the same) peer and that peer would dry run that test again.

Formal Execution

Formal execution is running a test or set of tests for the purpose of formal documentation. This means that the test results with the tester's name, date, and signature is kept for record, to be accessed at any time in the future. These records are usually summarized in a document and given to all stakeholders, so they can make a decision to release to software to the outside world (or not). Formal execution is not to be taken lightly! Integrity is of the utmost importance.

12 Bugs

Finding Bugs

A bug is a defect or anomaly in the software that doesn't function as expected, is inconsistent, or doesn't conform to a written requirement. Common synonyms include "defect", "issue", "problem", and "error".

When a bug is found, normally the finder locates a requirement that conflicts with the bug and writes a report before sending the report to a superior and or development. Occasionally a bug will not be related to a requirement, in which a bug report should still be created. Notice the example bugs below.

Example 1:
Requirement: The login button shall be blue at all times.
Implementation: In actuality, the tester notices that the login button is yellow. The software implementation conflicts with a written requirement, so this is most definitely a bug.

Example 2:
Requirement: No requirement is applicable.
Implementation: A mobile app login screen has an exit button that closes the app. After login, the mobile app has an exit button that logs the user out. The exit button does two different things, and therefore is inconsistent

behavior and should be filed as a bug.

Defect Clustering

When one bug is found, defect clustering should be considered. Defect clustering postulates that when one bug is found in a specific software feature, there is an increased likelihood that more exist in that same feature. For example, a desktop application has window that shows all active users. The users can be sorted by name, username, and registration date. A bug was just found in which the registration dates are sorted by day instead year. The tester should continue to test the sorting feature because there is an increased likelihood of finding more bugs in the same feature.

Reporting a Bug

Finding a software bug is incredibly rewarding, but also incredibly time consuming because of the reporting that must occur.

Whatever each company's policy is for reporting bugs, it usually takes a decent amount of time to clearly write exactly what happened, what was supposed to happen, and attach any visual evidence available.

Ensure before reporting a bug, that the tester does his or her due diligence to find and investigate all aspects of the bug. For example, make sure to test different software versions to see if the bug occurs in only one or all versions. Test to see if you can reproduce the bug doing different actions. Try to see if you can narrow the bug down to the exact place in software where it occurs. Lastly, check to see if there are any work-arounds to report.

For most clients, it takes me anywhere from ten to thirty minutes per bug report. At a minimum, each bug report should have the following information:

1. Summary: One to two sentences explaining what happened.
2. Expected Results: What should have happened.
3. Actual Results: What actually happened.
4. Steps to Reproduce: The exact steps another person can take to observe the same bug.
5. Requirements: The requirement tags of requirements that contradict the actual results.

Other information that should be included is any versions that were tested, what devices were used (if any), dates, times, and anything else that may be useful for a developer to investigate.

14 Test Activities

Many test activities are indicated below. There can be more activities if desired or established by a process.

Test Planning: Creating a documented test plan based on the software development plan that explains what to be tested for a particular version of software desired to be released. This document should be created as early as possible.

Requirements Review: Reviewing requirements to ensure clarity and testability.

Test Creation: Creating and drafting documented test cases.

Regression Testing: Testing partially or fully implemented features, released internally for testing. Also, re-testing software to ensure bugs are correctly fixed.

Reviewing Drafted Tests: Peer reviewing or inspecting drafted test cases for accuracy and requirements coverage.

Creating Traceability: Grouping and documenting requirement(s) with applicable test case(s).

Approving Tests: Marking reviewed test cases as

approved and ready for formal execution.

Dry Running Tests: Executing approved tests to obtain and report current software status.

Tool Validation: Making sure the independent software tools being used are adequate.

Formal Testing: Executing approved tests to formally document the status of a potential software release.

Test Debt: Analyzing current and past testing to document and fix and coverage gaps and issues with the test process. One example is to update the test process to have greater test efficiency, while still maintaining or increasing test quality.

Test Summary: Documenting the results of the formal testing.

15 Test Documentation

Test documentation can be daunting at first, but with practice, it becomes easier and achievable. From the tester's perspective, there are generally four specific deliverables that a test department is generally responsible for, and is required to provide for a given project. The deliverables are:

- Test Plan
- Test Protocol(s)
- Traceability
- Test Summary

Indirectly, testers may have some responsibility for the Software Requirements Specification but this is usually regarding review and possible approval of the document, not drafting or finalization.

The test plan is created at a very high level and should dictate what will be accomplished. All other referenced documents should never include test plan content. For example, a test protocol should never include any statements indicating that the test equipment should be referenced in the plan. It should always be the other way around, top-down. The plan should indicate what is in the test protocols, and what traceability should exist.

The test plan is usually drafted first and outlines many items, including references, software risks, features tested and not tested, pass/fail criteria, suspension and resumption criteria, test deliverables, scheduling, and staffing. The test plan provided by IEEE (829) is a great format and provides clear instructions, but may lack some required information for regulatory compliance..

The test plan should have a very high visibility from all test personnel so compliance is achieved. This allows many questions to be answered without debate or wasting time. Although specific scheduling with staffing and delegation is not determined by the plan (e.g., whom will test what feature), related information can be gleaned from the plan.

Keep in mind that the plan can and will most likely need be updated up until formal testing begins. Once finalized, the plan will need to be reviewed and approved by the powers that be.

After the test plan has at least been drafted, test cases are usually drafted as software requirements come into existence. Usually each test case is part of one or more protocols. A protocol is a document that contains one or more test cases. Each protocol must be reviewed and approved by appropriate approvers. For example, a test may be created by a tester, reviewed by a peer, and approved by the test lead or manager. In my experience, protocols are generally reviewed and approved right before formal testing to avoid multiple reviews and approvals.

Test traceability is of the utmost importance. It's needed not only for regulatory conformance, but also for test maintenance. Many types of traceability exist. For testing, the most applicable traceability is requirements to test cases. Traceability also includes system requirements to requirements, system requirements to test cases, risks to requirements, and risks to test cases. It is usually kept in one document, known as a traceability matrix (or traceability analysis), or in an appendix in the test summary although other implementations are acceptable.

For software testers, as previously mentioned, each test case should include what requirements it covers at a minimum in the test approach, but potentially also in each verification step. The traceability should also exist in a separate document (e.g., test summary).

Traceability is used by testers for three important reasons: to find testing gaps, regulatory/standards conformance, and to quickly find test case to execute or change. If there are risks or requirements that do not have related test cases in the same row, then likely more tests need to be created, or additional coverage needs to be added to an existing test.

Traceability is created in conjunction with test development. The test plan should dictate what traceability should be created. At a minimum, the requirements, test coverage, and risks need to be able to be traced to each other. The following is an example of traceability to be included for a product that has both system requirements and software requirements:

- Software Requirements to Test Cases
- System Requirements to Test Cases
- Risk Controls to Test Cases

The test summary can be drafted prior to formal testing, but may not be finished until after formal test is complete.

Lastly, the test summary should mirror the test plan and indicate if the test plan was adhered to, and if not, what wasn't. For example, if the test plan indicated that a "login" feature would be tested, then the test summary should indicate that the "login" feature was tested and summarize the test results. Like the other documents, this needs to be reviewed and approved. In addition, the test results are normally attached to an appendix of the test summary as well as noted in the test summary itself of what passed and failed. The test summary should also include what bugs still exist in the system.

16 Verification & Validation

Verification and validation are similar terms and can be confusing. Validation is ensuring that the user needs have been met. For example, a user needs requirement may state "A user shall be able to save their profile settings for subsequent visits". To validate that requirement, a test would change and save the profile settings, making sure that the requirement was satisfied. These user needs are generally tested through usability, (human factors) and user acceptance testing.

Verification is ensuring that the software requirements have been satisfied. For example, a software requirement may state "The software shall allow the user to change their password on the profile page". To verify, the tester would change the password on the profile page and verify it was saved.

There is little distinction between the two words, but the importance must be understood. When determining which term is correct for a given situation, consider the following:

- Verification (software requirement): Is the software right?

- Validation (user needs requirement): Do we have the right software?

17 Test Tools

Many test tools exist to aid and assist modern testers. The following criteria is listed by importance, and should be researched before implementing and purchasing any tools:

- How long does the tool take to setup and configure?
- How many people does it take to daily maintain the tool?
- Do you or the company have adequate resources to setup and maintain the tool?
- Have you tried using the tool?
- Would the tool save time and/or resources?
- How much does the tool cost?

Be cautious when trying tools. Not every tool will save time, and just because the tool is expensive does not mean that it is a good tool to use. Conversely, just because the tool is free does not meant that it is good either.

The most common testing tools can be divided into the following categories: test management, configuration management/source control, bug reporting, and test automation.

Test management tools aid in creating electronic cloud or network stored tests. Generally, these tools automatically create traceability between the test and requirements and provide a preconfigured test template. A test development process can usually be preconfigured as well, eliminating process confusion. Some of these tools also allow for paper-less test execution which can help speed up test execution time.

Configuration management and source control tools provide a way to save the history of any and all changes for each file, whether a document or code file. This enables developers to program features and merge their changes with everyone else's. This also allows for testing specific historical and non-released software versions.

Bug reporting tools allow for the display of preconfigured fields which allows a tester to save time filling out a bug report. Most tools also allow for the automatic distribution and assigning of bugs to specific developers.

Test automation tools allow tests to be created manually but executed automatically, usually with much quicker execution times and allowing for more comprehensive testing.

It's important to note that these tools should be validated at some level, otherwise, how do you know if the tool is functioning properly?

19 Software Development Lifecycle

Numerous types of software development lifecycles exist, and numerous variations of lifecycles exist for every major lifecycle. There are two main lifecycles that testers should be aware of: waterfall, and incremental (Agile being the most popular). No single development lifecycle is the correct one for all products and companies – each one must be custom-tailored to fit each company.

The waterfall software development lifecycle is strictly regulated and non-incremental, meaning that everything is completed in stages. For example, design review meetings are conducted until requirements are finalized. Once requirements are finalized, developers program the requirements. Once the developers feel that they are finished, they hand off the release candidate software to the test department which tests and files bugs. Once they file all their bugs, the software goes back to development until they fix almost everything, then the software goes back to test. This process repeats through until the software is released. The testers have to create tests based on what they think the software will look like and behave like based on the finalized requirements, and will have to wait to debug the tests until they receive a working version of the software.

The incremental software development model

attempts to release software quicker by creating requirements for one function or feature at a time. After the requirement(s) are agreed upon, developers code the requirement and testers create the test. Usually daily or weekly versions of the software are released which allow testers to work more in parallel with developers and file bug reports quicker.

20 Test Automation

Introduction

Test automation has increased dramatically over the past years and maintains its supremacy as a popular topic amongst software testers. Only recently have test automation tools become advanced enough to allow for test benefits, such as saving time on execution and regression testing. Although there can be significant benefits from test automation, it's important to note that most bugs are found by ad-hoc testing. Therefore, automation testing should never fully replace manual testing, but rather make testing more efficient in order for software testers to ad-hoc test software. There are two main types of test automation tools available: programmed, and capture/playback. There are also tools with any combination of the two tools. Capture/playback tools attempt to allow software testers without programming experience the ability to create automated tests whereas programmed automated tests need each software tester to have at least some programming experience.

Concepts

Aliasing is one of the most important concepts for efficiency in automation. Aliasing is masking a mapped object's path with a variable name so the path is only referenced in one place. All other references to that object use the variable name, so when an object's path changes due to software updates, only one reference needs to be updated in the automated testing rather than updating fifty or more references.

Data-driven (iterative) testing is another important concept available through test automation. Something that could take a manual tester ten or more hours to complete could take an automated tester a mere hour depending on the tool used. Data-driven testing loops through a test a specified number of times. For example, to test a login username field, a tester can automate this test for any number of invalid user names sometimes quicker than manually testing can.

Regression testing is one of biggest benefits of test automation. Each test should only be created once, and test execution time is normally greatly reduced when compared with manual execution time. Therefore, when the same test needs to be re-executed, it can be with the same quick execution time while the software tester's time is reserved for additional tasks.

Tools

As previously stated, it's never good to judge a tool by the price. There are ten-thousand-dollar test automation tools that should not be touched. I experienced one such tool with a company that had four full-time employees maintaining the tool, and the tool still lacked important and needed features. There are also dozens of free test automation tools that require weeks of configuration and setup, and also require at least one full-time person to manage the new versions, updates, and other 3rd party tools needed for that one free tool. All the time spent configuring and maintaining actually ends up costing more money than it would purchasing a decent tool that works out-of-the-box.

The best test automation tools should adhere to the test tool criteria, should easily provide test automation concepts for the tester, and should also allow for non-programmers to develop tests. All the above should be out-of-the-box, when possible.

Glossary

Definitions within are from FDA's Global Principals of Software Validation, and:

American Software Testing Qualifications Board: Standard Glossary of Terms used in Software Testing; version 3.01, 2015.

Acceptance criteria [ASTQB] The exit criteria that a component or system must satisfy in order to be accepted by a user, customer, or other authorized entity.

Acceptance testing [ASTQB]*(See Also: user acceptance testing)* Formal testing with respect to user needs, requirements, and business processes conducted to determine whether or not a system satisfies the acceptance criteria and to enable the user, customers or other authorized entity to determine whether or not to accept the system.

Accessibility testing [ASTQB] Testing to determine the ease by which users with disabilities can use a component or system.

Accuracy [ASTQB]*(See Also: functionality)* The capability of the software product to provide the right or agreed results or effects with the needed degree of precision.

Accuracy testing [ASTQB]*(See Also: accuracy)* Testing to determine the accuracy of a software product.

Actor [ASTQB]User or any other person or system that interacts with the system under test in a specific way.

Actual result [ASTQB]*(Synonyms: actual outcome)* The behavior produced/observed when a component or system is tested.

Ad hoc testing [ASTQB]Testing carried out informally. No formal test preparation takes place, no recognized test design technique is used, there are no expectations for results and arbitrariness guides the test execution activity.

Adaptability [ASTQB]*(See Also: portability)* The capability of the software product to be adapted for different specified environments without applying actions or means other than those provided for this purpose for the software considered.

Agile Manifesto [ASTQB] A statement on the values that underpin Agile software development. The values are: individuals and interactions over processes and tools, responding to change over following a plan, customer collaboration over contract negotiation, working software over comprehensive documentation.

Agile software development [ASTQB] A group of software development methodologies based on iterative incremental development, where requirements and solutions evolve through collaboration between self-organizing cross-functional teams.

Agile testing [ASTQB]*(See Also: test-driven development)* Testing practice for a project using Agile software development methodologies, incorporating techniques and methods, such as extreme programming (XP), treating development as the customer of testing and emphasizing the test-first design paradigm.

Alpha testing [ASTQB]Simulated or actual operational testing by potential users/customers or an independent test team at the developers' site, but outside the development organization. Alpha testing is often employed for off-the-shelf software as a form of internal acceptance testing.

Analytical test strategy [ASTQB] A test strategy whereby the test team analyzes the test basis to identify the test conditions to cover.

Analytical testing [ASTQB] Testing based on a systematic analysis of e.g., product risks or requirements.

Analyzability [ASTQB]*(See Also: maintainability)* The capability of the software product to be diagnosed for deficiencies or causes of failures in the software, or for the parts to be modified to be identified.

Anomaly [ASTQB]*(See Also: defect, error, fault, failure, incident, problem)* Any condition that deviates from expectation based on requirements specifications, design documents, user documents, standards, etc., or from someone's perception or experience. Anomalies may be found during, but not limited to, reviewing, testing, analysis, compilation, or use of software products or applicable documentation.

Anti-pattern [ASTQB] Repeated action, process, structure or reusable solution that initially appears to be beneficial and is commonly used but is ineffective and/or counterproductive in practice.

API [ASTQB] Acronym for Application Programming Interface.

API testing [ASTQB] Testing performed by submitting commands to the software under test using programming interfaces of the application directly.

Assessment report [ASTQB]*(See Also: process assessment)* A document summarizing the assessment results, e.g., conclusions, recommendations and findings.

Assessor [ASTQB] A person who conducts an assessment. Any member of an assessment team.

Atomic condition [ASTQB] A condition that cannot be decomposed, i.e., a condition that does not contain two or more single conditions joined by a logical operator (AND, OR, XOR).

Attack-based testing [ASTQB]*(See Also: attack)* An experience-based testing technique that uses software attacks to induce failures, particularly security related failures.

Attractiveness [ASTQB](*See Also: usability)* The capability of the software product to be attractive to the user.

Audit [ASTQB] An independent evaluation of software products or processes to ascertain compliance to standards, guidelines, specifications, and/or procedures based on objective criteria, including documents that specify: (1) the form or content of the products to be produced, (2) the process by which the products shall be produced, (3) how compliance to standards or guidelines shall be measured.

Audit trail [ASTQB] A path by which the original input to a process (e.g., data) can be traced back through the process, taking the process output as a starting point. This facilitates defect analysis and allows a process audit to be carried out.

Automated testware [ASTQB] Testware used in automated testing, such as tool scripts.

Automation code defect density [ASTQB](*See Also: defect density)* Defect density of a component of the test automation code.

Availability [ASTQB] The degree to which a component or system is operational and accessible when required for use. Often expressed as a percentage.

Balanced scorecard [ASTQB](*See Also: corporate dashboard, scorecard)* A strategic tool for measuring whether the operational activities of a company are aligned with its objectives in terms of business vision and strategy.

Baseline [ASTQB] A specification or software product that has been formally reviewed or agreed upon, that thereafter serves as the basis for further development, and that can be changed only through a formal change control process.

Basic block [ASTQB] A sequence of one or more consecutive executable statements containing no branches. Note: A node in a control flow graph represents a basic block.

Basis test set [ASTQB] A set of test cases derived from the internal structure of a component or specification to ensure that 100% of a specified coverage criterion will be achieved.

Behavior [ASTQB] The response of a component or system to a set of input values and preconditions.

Benchmark test [ASTQB] (1) A standard against which measurements or comparisons can be made. (2) A test that is used to compare components or systems to each other or to a standard as in (1).

Best practice [ASTQB] A superior method or innovative practice that contributes to the improved performance of an organization under given context, usually recognized as "best" by other peer organizations.

Beta testing [ASTQB](*Synonyms: field testing*) Operational testing by potential and/or existing users/customers at an external site not otherwise involved with the developers, to determine whether or not a component or system satisfies the user/customer needs and fits within the business processes. Beta testing is often employed as a form of external acceptance testing for off-the-shelf software in order to acquire feedback from the market.

Big-bang testing [ASTQB](*See Also: integration testing*) An integration testing approach in which software elements, hardware elements, or both are combined all at once into a component or an overall system, rather than in stages.

Black-box test design technique [ASTQB](*Synonyms: black-box technique, specification-based technique, specification-based test design technique*) Procedure to derive and/or select test cases based on an analysis of the specification, either functional or non-functional, of a component or system without reference to its internal structure.

Black-box testing [ASTQB] *Synonyms: specification-based testing* Testing, either functional or non-functional, without reference to the internal structure of the component or system.

Blocked test case [ASTQB] A test case that cannot be executed because the preconditions for its execution are not fulfilled.

Bottom-up testing [ASTQB](*See Also: integration testing*) An incremental approach to integration testing where the lowest level components are tested first, and then used to facilitate the testing of higher level components. This process is repeated until the component at the top of the hierarchy is tested.

Boundary value [ASTQB] An input value or output value which is on the edge of an equivalence partition or at the smallest incremental distance on either side of an edge, for example the minimum or maximum value of a range.

Boundary value analysis [ASTQB](*See Also: boundary value*) A black-box test design technique in which test cases are designed based on boundary values.

Boundary value coverage [ASTQB] The percentage of boundary values that have been exercised by a test suite.

Branch [ASTQB] A basic block that can be selected for execution based on a program construct in which one of two or more alternative program paths is available, e.g., case, jump, go to, if-then-else.

Branch coverage [ASTQB] The percentage of branches that have been exercised by a test suite. 100% branch coverage implies both 100% decision coverage and 100% statement coverage.

Branch testing [ASTQB] A white-box test design technique in which test cases are designed to execute branches.

Buffer [ASTQB] A device or storage area used to store data temporarily for differences in rates of data flow, time or occurrence of events, or amounts of data that can be handled by the devices or processes involved in the transfer or use of the data.

Buffer overflow [ASTQB](*See Also: buffer)* A memory access failure due to the attempt by a process to store data beyond the boundaries of a fixed length buffer, resulting in overwriting of adjacent memory areas or the raising of an overflow exception.

Build verification test (BVT) [ASTQB](*See Also: regression testing, smoke test)* A set of automated tests which validates the integrity of each new build and verifies its key/core functionality, stability and testability. It is an industry practice when a high frequency of build releases occurs (e.g., Agile projects) and it is run on every new build before the build is released for further testing.

Burndown chart [ASTQB] A publicly displayed chart that depicts the outstanding effort versus time in an iteration. It shows the status and trend of completing the tasks of the iteration. The X-axis typically represents days in the sprint, while the Y-axis is the remaining effort (usually either in ideal engineering hours or story points).

Business process-based testing [ASTQB] An approach to testing in which test cases are designed based on descriptions and/or knowledge of business processes.

Call graph [ASTQB] An abstract representation of calling relationships between subroutines in a program.

Capability Maturity Model Integration (CMMI) [ASTQB] A framework that describes the key elements of an effective product development and maintenance process. The Capability Maturity Model Integration covers best-practices for planning, engineering and managing product development and maintenance.

Capture/playback [ASTQB] A test automation approach, where inputs to the test object are recorded during manual testing in order to generate automated test scripts that could be executed later (i.e. replayed).

Capture/playback tool [ASTQB](*Synonyms: capture/replay tool, record/playback tool)* A type of test execution tool where inputs are recorded during manual testing in order to generate automated test scripts that can be executed later (i.e. replayed). These tools are often used to support automated regression testing.

CASE [ASTQB] Acronym for Computer Aided Software Engineering.

CAST [ASTQB](*See Also: test automation)* Acronym for Computer Aided Software Testing.

Causal analysis [ASTQB] The analysis of defects to determine their root cause.

Cause-effect diagram [ASTQB] A graphical representation used to organize and display the interrelationships of various possible root causes of a problem. Possible causes of a real or potential defect or failure are organized in categories and subcategories in a horizontal tree-structure, with the (potential) defect or failure as the root node.

Cause-effect graph [ASTQB] A graphical representation of inputs and/or stimuli (causes) with their associated outputs (effects), which can be used to design test cases.

Cause-effect graphing [ASTQB](*Synonyms: cause-effect analysis)* A black-box test design technique in which test cases are designed from cause-effect graphs.

Certification [ASTQB] The process of confirming that a component, system or person complies with its specified requirements, e.g., by passing an exam.

Change management [ASTQB](*See Also: configuration management)* (1) A structured approach to transitioning individuals and organizations from a current state to a desired future state. (2) Controlled way to effect a change, or a proposed change, to a product or service.

Changeability [ASTQB](*See Also: maintainability)* The capability of the software product to enable specified modifications to be implemented.

Checklist-based testing [ASTQB] An experience-based test design technique whereby the experienced tester uses a high-level list of items to be noted, checked, or remembered, or a set of rules or criteria against which a product has to be verified.

Classification tree [ASTQB](*See Also: classification tree method)* A tree showing equivalence partitions hierarchically ordered, which is used to design test cases in the classification tree method.

Classification tree method [ASTQB](*See Also: combinatorial testing)* A black-box test design technique in which test cases, described by means of a classification tree, are designed to execute combinations of representatives of input and/or output domains.

CLI [ASTQB] *Acronym for Command-Line Interface.*

CLI testing [ASTQB] Testing performed by submitting commands to the software under test using a dedicated command-line interface.

Co-existence [ASTQB]*(See Also: portability)* The capability of the software product to co-exist with other independent software in a common environment sharing common resources.

Code [ASTQB] Computer instructions and data definitions expressed in a programming language or in a form output by an assembler, compiler or other translator.

Code coverage [ASTQB] An analysis method that determines which parts of the software have been executed (covered) by the test suite and which parts have not been executed, e.g., statement coverage, decision coverage or condition coverage.

Codependent behavior [ASTQB] Excessive emotional or psychological dependence on another person, specifically in trying to change that person's current (undesirable) behavior while supporting them in continuing that behavior. For example, in software testing, complaining about late delivery to test and yet enjoying the necessary "heroism", working additional hours to make up time when delivery is running late, therefore reinforcing the lateness.

Combinatorial testing [ASTQB]*(See Also: classification tree method, n-wise testing, pairwise testing, orthogonal array testing)* A black-box test design technique in which test cases are designed to execute specific combinations of values of several parameters.

Commercial off-the-shelf (COTS)
[ASTQB]*(Synonyms: off-the-shelf software)* A software product that is developed for the general market, i.e. for a large number of customers, and that is delivered to many customers in identical format.

Complaint [FDA 820.3] Any written, electronic, or oral communication that alleges deficiencies related to the identity, quality, durability, reliability, safety, effectiveness, or performance of a device after it is released for distribution.

Compiler [ASTQB] A software tool that translates programs expressed in a high-order language into their machine language equivalents.

Complexity [ASTQB](*See Also: cyclomatic complexity)* The degree to which a component or system has a design and/or internal structure that is difficult to understand, maintain and verify.

Compliance [ASTQB] The capability of the software product to adhere to standards, conventions or regulations in laws and similar prescriptions.

Compliance testing [ASTQB](*Synonyms: conformance testing, regulation testing, standards testing)* Testing to determine the compliance of the component or system.

Component [ASTQB](*Synonyms: module, unit)*(1) A minimal software item that can be tested in isolation. [FDA 820.3](2) Any raw material, substance, piece, part, software, firmware, labeling, or assembly which is intended to be included as part of the finished, packaged, and labeled device.

Component integration testing [ASTQB](*Synonyms: link testing)* Testing performed to expose defects in the interfaces and interaction between integrated components.

Component specification [ASTQB] A description of a component's function in terms of its output values for specified input values under specified conditions, and required non-functional behavior (e.g., resource-utilization).

Component testing [ASTQB](*Synonyms: module testing, program testing, unit testing*) The testing of individual software components.

Compound condition [ASTQB](*Synonyms: multiple condition*) Two or more single conditions joined by means of a logical operator (AND, OR or XOR), e.g., A>B AND C>1000.

Concurrency testing [ASTQB] Testing to determine how the occurrence of two or more activities within the same interval of time, achieved either by interleaving the activities or by simultaneous execution, is handled by the component or system.

Condition [ASTQB](*See Also: condition testing Synonyms: branch condition*) A logical expression that can be evaluated as True or False, e.g., A>B.

Condition coverage [ASTQB](*Synonyms: branch condition coverage*) The percentage of condition outcomes that have been exercised by a test suite. 100% condition coverage requires each single condition in every decision statement to be tested as True and False.

Condition outcome [ASTQB] The evaluation of a condition to True or False.

Condition testing [ASTQB] A white-box test design technique in which test cases are designed to execute condition outcomes.

Confidence interval [ASTQB] In managing project risks, the period of time within which a contingency action must be implemented in order to be effective in reducing the impact of the risk.

Configuration [ASTQB] The composition of a component or system as defined by the number, nature, and interconnections of its constituent parts.

Configuration auditing [ASTQB] The function to check on the contents of libraries of configuration items, e.g., for standards compliance.

Configuration control [ASTQB](*Synonyms: change control, version control)* An element of configuration management, consisting of the evaluation, coordination, approval or disapproval, and implementation of changes to configuration items after formal establishment of their configuration identification.

Configuration control board (CCB) [ASTQB](*Synonyms: change control board)* A group of people responsible for evaluating and approving or disapproving proposed changes to configuration items, and for ensuring implementation of approved changes.

Configuration identification [ASTQB] An element of configuration management, consisting of selecting the configuration items for a system and recording their functional and physical characteristics in technical documentation.

Configuration item [ASTQB] An aggregation of hardware, software or both, that is designated for configuration management and treated as a single entity in the configuration management process.

Configuration management [ASTQB] A discipline applying technical and administrative direction and surveillance to identify and document the functional and physical characteristics of a configuration item, control changes to those characteristics, record and report change processing and implementation status, and verify compliance with specified requirements.

Configuration management tool [ASTQB] A tool that provides support for the identification and control of configuration items, their status over changes and versions, and the release of baselines consisting of configuration items.

Confirmation testing [ASTQB](*Synonyms: re-testing)* Testing that runs test cases that failed the last time they were run, in order to verify the success of corrective actions.

Consultative test strategy [ASTQB] A test strategy whereby the test team relies on the input of one or more key stakeholders to determine the details of the strategy.

Consultative testing [ASTQB] Testing driven by the advice and guidance of appropriate experts from outside the test team (e.g., technology experts and/or business domain experts).

Content-based model [ASTQB]*(Synonyms: content reference model)* A process model providing a detailed description of good engineering practices, e.g., test practices.

Continuous representation [ASTQB] A capability maturity model structure wherein capability levels provide a recommended order for approaching process improvement within specified process areas.

Control chart [ASTQB]*(Synonyms: Shewhart chart)* A statistical process control tool used to monitor a process and determine whether it is statistically controlled. It graphically depicts the average value and the upper and lower control limits (the highest and lowest values) of a process.

Control flow [ASTQB] A sequence of events (paths) in the execution through a component or system.

Control flow analysis [ASTQB] A form of static analysis based on a representation of unique paths (sequences of events) in the execution through a component or system. Control flow analysis evaluates the integrity of control flow structures, looking for possible control flow anomalies such as closed loops or logically unreachable process steps.

Control flow graph [ASTQB] An abstract representation of all possible sequences of events (paths) in the execution through a component or system.

Control flow testing [ASTQB](*See Also: decision testing, condition testing, path testing*) An approach to structure-based testing in which test cases are designed to execute specific sequences of events. Various techniques exist for control flow testing, e.g., decision testing, condition testing, and path testing, that each have their specific approach and level of control flow coverage.

Control number [FDA 820.3] any distinctive symbols, such as a distinctive combination of letters or numbers, or both, from which the history of the manufacturing, packaging, labeling, and distribution of a unit, lot, or batch of finished devices can be determined.

Convergence metric [ASTQB] A metric that shows progress toward a defined criterion, e.g., convergence of the total number of tests executed to the total number of tests planned for execution.

Conversion testing [ASTQB](*Synonyms: migration testing*) Testing of software used to convert data from existing systems for use in replacement systems.

Corporate dashboard [ASTQB](*See Also: balanced scorecard, dashboard*) A dashboard-style representation of the status of corporate performance data.

Cost of quality [ASTQB] The total costs incurred on quality activities and issues and often split into prevention costs, appraisal costs, internal failure costs and external failure costs.

Coverage [ASTQB](*Synonyms: test coverage*) The degree, expressed as a percentage, to which a specified coverage item has been exercised by a test suite.

Coverage analysis [ASTQB] Measurement of achieved coverage to a specified coverage item during test execution referring to predetermined criteria to determine whether additional testing is required and if so, which test cases are needed.

Coverage item [ASTQB] An entity or property used as a basis for test coverage, e.g., equivalence partitions or code statements.

Coverage tool [ASTQB]*(Synonyms: coverage measurement tool)* A tool that provides objective measures of what structural elements, e.g., statements, branches have been exercised by a test suite.

Critical success factor [ASTQB] An element necessary for an organization or project to achieve its mission. Critical success factors are the critical factors or activities required for ensuring the success.

Critical Testing Processes (CTP) [ASTQB]*(See Also: content-based model)* A content-based model for test process improvement built around twelve critical processes. These include highly visible processes, by which peers and management judge competence and mission-critical processes in which performance affects the company's profits and reputation.

Custom software [ASTQB]*(Synonyms: bespoke software)* Software developed specifically for a set of users or customers. The opposite is off-the-shelf software.

Custom tool [ASTQB] A software tool developed specifically for a set of users or customers.

Cyclomatic complexity [ASTQB]*(Synonyms: cyclomatic number)* The maximum number of linear, independent paths through a program. Cyclomatic complexity may be computed as $L = N + 2P$, where L = the number of edges/links in a graph, N = the number of nodes in a graph, P = the number of disconnected parts of the graph (e.g., a called graph or subroutine).

Daily build [ASTQB] A development activity whereby a complete system is compiled and linked every day (often overnight), so that a consistent system is available at any time including all latest changes.

Dashboard [ASTQB](*See Also: corporate dashboard, scorecard*) A representation of dynamic measurements of operational performance for some organization or activity, using metrics represented via metaphors such as visual dials, counters, and other devices resembling those on the dashboard of an automobile, so that the effects of events or activities can be easily understood and related to operational goals.

Data definition [ASTQB] An executable statement where a variable is assigned a value.

Data flow [ASTQB] An abstract representation of the sequence and possible changes of the state of data objects, where the state of an object is any of creation, usage, or destruction.

Data flow analysis [ASTQB] A form of static analysis based on the definition and usage of variables.

Data flow coverage [ASTQB] The percentage of definition-use pairs that have been exercised by a test suite.

Data flow testing [ASTQB] A white-box test design technique in which test cases are designed to execute definition-use pairs of variables.

Data quality [ASTQB] An attribute of data that indicates correctness with respect to some pre-defined criteria, e.g., business expectations, requirements on data integrity, data consistency.

Data-driven testing [ASTQB](*See Also: keyword-driven testing*) A scripting technique that stores test input and expected results in a table or spreadsheet, so that a single control script can execute all of the tests in the table. Data-driven testing is often used to support the application of test execution tools such as capture/playback tools.

Database integrity testing [ASTQB] Testing the methods and processes used to access and manage the data(base), to ensure access methods, processes and data rules function as expected and that during access to the database, data is not corrupted or unexpectedly deleted, updated or created.

Dd-path [ASTQB](*See Also: path*) A path between two decisions of an algorithm, or two decision nodes of a corresponding graph, that includes no other decisions.

Debugging [ASTQB] The process of finding, analyzing and removing the causes of failures in software.

Debugging tool [ASTQB](*Synonyms: debugger*) A tool used by programmers to reproduce failures, investigate the state of programs and find the corresponding defect. Debuggers enable programmers to execute programs step by step, to halt a program at any program statement and to set and examine program variables.

Decision [ASTQB] A program point at which the control flow has two or more alternative routes. A node with two or more links to separate branches.

Decision condition coverage [ASTQB] The percentage of all condition outcomes and decision outcomes that have been exercised by a test suite. 100% decision condition coverage implies both 100% condition coverage and 100% decision coverage.

Decision condition testing [ASTQB] A white-box test design technique in which test cases are designed to execute condition outcomes and decision outcomes.

Decision coverage [ASTQB] The percentage of decision outcomes that have been exercised by a test suite. 100% decision coverage implies both 100% branch coverage and 100% statement coverage.

Decision outcome [ASTQB] The result of a decision (which therefore determines the branches to be taken).

Decision table [ASTQB](*Synonyms: cause-effect decision table*) A table showing combinations of inputs and/or stimuli (causes) with their associated outputs and/or actions (effects), which can be used to design test cases.

Decision table testing [ASTQB](*See Also: decision table*) A black-box test design technique in which test cases are designed to execute the combinations of inputs and/or stimuli (causes) shown in a decision table.

Decision testing [ASTQB] A white-box test design technique in which test cases are designed to execute decision outcomes.

Defect [ASTQB](*Synonyms: bug, fault, problem*) A flaw in a component or system that can cause the component or system to fail to perform its required function, e.g., an incorrect statement or data definition. A defect, if encountered during execution, may cause a failure of the component or system.

Defect density [ASTQB](*Synonyms: fault density*) The number of defects identified in a component or system divided by the size of the component or system (expressed in standard measurement terms, e.g., lines-of-code, number of classes or function points).

Defect Detection Percentage (DDP) [ASTQB](*See Also: escaped defects Synonyms: Fault Detection Percentage (FDP)*) The number of defects found by a test level, divided by the number found by that test level and any other means afterwards.

Defect management [ASTQB](*Synonyms: problem management*) The process of recognizing, investigating, taking action and disposing of defects. It involves recording defects, classifying them and identifying the impact.

Defect management committee [ASTQB](*Synonyms: defect triage committee*) A cross-functional team of stakeholders who manage reported defects from initial detection to ultimate resolution (defect removal, defect deferral, or report cancellation). In some cases, the same team as the configuration control board.

Defect management tool [ASTQB]*(See Also: incident management tool)(Synonyms: bug tracking tool, defect tracking tool)* A tool that facilitates the recording and status tracking of defects and changes. They often have workflow-oriented facilities to track and control the allocation, correction and re-testing of defects and provide reporting facilities.

Defect masking [ASTQB]*(Synonyms: fault masking)* An occurrence in which one defect prevents the detection of another.

Defect report [ASTQB]*(Synonyms: bug report, problem report)* A document reporting on any flaw in a component or system that can cause the component or system to fail to perform its required function.

Defect taxonomy [ASTQB] *(Synonyms: bug taxonomy)* A system of (hierarchical) categories designed to be a useful aid for reproducibly classifying defects.

Defect type [ASTQB]*(Synonyms: defect category)* An element in a taxonomy of defects. Defect taxonomies can be identified with respect to a variety of considerations, including, but not limited to: Phase or development activity in which the defect is created, e.g., a specification error or a coding error, Characterization of defects, e.g., an "off-by-one" defect, Incorrectness, e.g., an incorrect relational operator, a programming language syntax error, or an invalid assumption, Performance issues, e.g., excessive execution time, insufficient availability.

Defect-based test design technique [ASTQB]*(See Also: defect taxonomy Synonyms: defect-based technique)* A procedure to derive and/or select test cases targeted at one or more defect types, with tests being developed from what is known about the specific defect type.

Definition-use pair [ASTQB] The association of a definition of a variable with the subsequent use of that variable. Variable uses include computational (e.g., multiplication) or to direct the execution of a path (predicate use).

Deliverable [ASTQB] Any (work) product that must be delivered to someone other than the (work) product's author.

Deming cycle [ASTQB] An iterative four-step problem-solving process (plan-do-check-act) typically used in process improvement.

Design history file (DHF) [FDA 820.3] A compilation of records which describes the design history of a finished device.

Design input [FDA 820.3] the physical and performance requirements of a device that are used as a basis for device design.

Design output [FDA 820.3] the results of a design effort at each design phase and at the end of the total design effort. The finished design output is the basis for the device master record. The total finished design output consists of the device, its packaging and labeling, and the device master record.

Design review [FDA 820.3] a documented, comprehensive, systematic examination of a design to evaluate the adequacy of the design requirements, to evaluate the capability of the design to meet these requirements, and to identify problems.

Design-based testing [ASTQB] An approach to testing in which test cases are designed based on the architecture and/or detailed design of a component or system (e.g., tests of interfaces between components or systems).

Desk checking [ASTQB](*See Also: static testing*) Testing of software or a specification by manual simulation of its execution.

Development testing [ASTQB] Formal or informal testing conducted during the implementation of a component or system, usually in the development environment by developers.

Device history record (DHR) [FDA 820.3] A compilation of records containing the production history of a finished device.

Device master record (DMR) [FDA 820.3] A compilation of records containing the procedures and specifications for a finished device.

Documentation testing [ASTQB] Testing the quality of the documentation, e.g., user guide or installation guide.

Domain [ASTQB] The set from which valid input and/or output values can be selected.

Domain analysis [ASTQB](*See Also: boundary value analysis, equivalence partitioning*) A black-box test design technique that is used to identify efficient and effective test cases when multiple variables can or should be tested together. It builds on and generalizes equivalence partitioning and boundary values analysis.

Driver [ASTQB] *Synonyms: test driver* A software component or test tool that replaces a component that takes care of the control and/or the calling of a component or system.

Dynamic analysis [ASTQB] The process of evaluating behavior, e.g., memory performance, CPU usage, of a system or component during execution.

Dynamic analysis tool [ASTQB] A tool that provides run-time information on the state of the software code. These tools are most commonly used to identify unassigned pointers, check pointer arithmetic and to monitor the allocation, use and de-allocation of memory and to flag memory leaks.

Dynamic comparison [ASTQB] Comparison of actual and expected results, performed while the software is being executed, for example by a test execution tool.

Dynamic testing [ASTQB] Testing that involves the execution of the software of a component or system.

Effectiveness [ASTQB]*(See Also: efficiency)* The capability of producing an intended result.

Efficiency [ASTQB] (1) The capability of the software product to provide appropriate performance, relative to the amount of resources used, under stated conditions. (2) The capability of a process to produce the intended outcome, relative to the amount of resources used.

Efficiency testing [ASTQB] Testing to determine the efficiency of a software product.

Elementary comparison testing [ASTQB] A black-box test design technique in which test cases are designed to execute combinations of inputs using the concept of modified condition decision coverage.

Embedded iterative model [ASTQB] A development lifecycle sub-model that applies an iterative approach to detailed design, coding and testing within an overall sequential model. In this case, the high-level design documents are prepared and approved for the entire project but the actual detailed design, code development and testing are conducted in iterations.

Emotional intelligence [ASTQB] The ability, capacity, and skill to identify, assess, and manage the emotions of one's self, of others, and of groups.

Emulator [ASTQB]*(See Also: simulator)* A device, computer program, or system that accepts the same inputs and produces the same outputs as a given system.

Entry criteria [ASTQB] The set of generic and specific conditions for permitting a process to go forward with a defined task, e.g., test phase. The purpose of entry criteria is to prevent a task from starting which would entail more (wasted) effort compared to the effort needed to remove the failed entry criteria.

Entry point [ASTQB] An executable statement or process step which defines a point at which a given process is intended to begin.

Equivalence partition [ASTQB](*Synonyms: equivalence class)* A portion of an input or output domain for which the behavior of a component or system is assumed to be the same, based on the specification.

Equivalence partition coverage [ASTQB] The percentage of equivalence partitions that have been exercised by a test suite.

Equivalence partitioning [ASTQB](*Synonyms: partition testing)* A black-box test design technique in which test cases are designed to execute representatives from equivalence partitions. In principle, test cases are designed to cover each partition at least once.

Equivalent manual test effort (EMTE) [ASTQB] Effort required for running tests manually.

Error [ASTQB] (*Synonyms: mistake)* A human action that produces an incorrect result.

Error guessing [ASTQB] A test design technique where the experience of the tester is used to anticipate what defects might be present in the component or system under test as a result of errors made, and to design tests specifically to expose them.

Error tolerance [ASTQB] The ability of a system or component to continue normal operation despite the presence of erroneous inputs.

Escaped defect [ASTQB](*See Also: Defect Detection Percentage)* A defect that was not detected in a previous test level which is supposed to find such type of defects.

Establish [FDA 820.3] To define, document (in writing or electronically), and implement.

Exception handling [ASTQB] Behavior of a component or system in response to erroneous input, from either a human user or from another component or system, or to an internal failure.

Executable statement [ASTQB] A statement which, when compiled, is translated into object code, and which will be executed procedurally when the program is running and may perform an action on data.

Exercised [ASTQB] A program element is said to be exercised by a test case when the input value causes the execution of that element, such as a statement, decision, or other structural element.

Exhaustive testing [ASTQB](*Synonyms: complete testing*) A test approach in which the test suite comprises all combinations of input values and preconditions.

Exit criteria [ASTQB](*Synonyms: completion criteria, test completion criteria*) The set of generic and specific conditions, agreed upon with the stakeholders for permitting a process to be officially completed. The purpose of exit criteria is to prevent a task from being considered completed when there are still outstanding parts of the task which have not been finished. Exit criteria are used to report against and to plan when to stop testing.

Exit point [ASTQB] An executable statement or process step which defines a point at which a given process is intended to cease.

Expected result [ASTQB] (*Synonyms: expected outcome, predicted outcome*) The behavior predicted by the specification, or another source, of the component or system under specified conditions.

Experience-based test design technique
[ASTQB](*Synonyms: experience-based technique)*
Procedure to derive and/or select test cases based on
the tester's experience, knowledge and intuition.

Experience-based testing [ASTQB] Testing based on
the tester's experience, knowledge and intuition.

Exploratory testing [ASTQB] An informal test design
technique where the tester actively controls the design
of the tests as those tests are performed and uses
information gained while testing to design new and
better tests.

Extreme Programming (XP) [ASTQB](*See Also: Agile
software development)* A software engineering
methodology used within Agile software development
whereby core practices are programming in pairs,
doing extensive code review, unit testing of all code,
and simplicity and clarity in code.

Factory acceptance testing [ASTQB](*See Also: alpha
testing)* Acceptance testing conducted at the site at
which the product is developed and performed by
employees of the supplier organization, to determine
whether or not a component or system satisfies the
requirements, normally including hardware as well as
software.

Fail [ASTQB](*Synonyms: test fail)* A test is deemed to
fail if its actual result does not match its expected
result.

Failover testing [ASTQB](*See Also: recoverability
testing)* Testing by simulating failure modes or
actually causing failures in a controlled environment.
Following a failure, the failover mechanism is tested to
ensure that data is not lost or corrupted and that any
agreed service levels are maintained (e.g., function
availability or response times).

Failure [ASTQB] Deviation of the component or system
from its expected delivery, service or result.

Failure mode [ASTQB] The physical or functional manifestation of a failure. For example, a system in failure mode may be characterized by slow operation, incorrect outputs, or complete termination of execution.

Failure Mode and Effect Analysis (FMEA) [ASTQB]*(See Also: Failure Mode, Effect and Criticality Analysis Synonyms: Software Failure Mode and Effect Analysis)* A systematic approach to risk identification and analysis of identifying possible modes of failure and attempting to prevent their occurrence.

Failure Mode, Effects, and Criticality Analysis (FMECA) [ASTQB] *(See Also: Failure Mode and Effect Analysis Synonyms: software failure mode)* An extension of FMEA, as in addition to the basic FMEA, it includes a criticality analysis, which is used to chart the probability of failure modes against the severity of their consequences. The result highlights failure modes with relatively high probability and severity of consequences, allowing remedial effort to be directed where it will produce the greatest value.

Failure rate [ASTQB] The ratio of the number of failures of a given category to a given unit of measure, e.g., failures per unit of time, failures per number of transactions, failures per number of computer runs.

False-negative result [ASTQB]*(Synonyms: false-pass result)* A test result which fails to identify the presence of a defect that is actually present in the test object.

False-positive result [ASTQB]*(Synonyms: false-fail result)* A test result in which a defect is reported although no such defect actually exists in the test object.

Fault attack [ASTQB]*(See Also: negative testing Synonyms: attack)* Directed and focused attempt to evaluate the quality, especially reliability, of a test object by attempting to force specific failures to occur.

Fault injection [ASTQB](*See Also: fault tolerance)* The process of intentionally adding defects to a system for the purpose of finding out whether the system can detect, and possibly recover from, a defect. Fault injection is intended to mimic failures that might occur in the field.

Fault seeding [ASTQB](*Synonyms: bebugging, error seeding)* The process of intentionally adding defects to those already in the component or system for the purpose of monitoring the rate of detection and removal, and estimating the number of remaining defects. Fault seeding is typically part of development (pre-release) testing and can be performed at any test level (component, integration, or system).

Fault seeding tool [ASTQB](*Synonyms: error seeding tool)* A tool for seeding (i.e., intentionally inserting) faults in a component or system.

Fault tolerance [ASTQB](*See Also: reliability, robustness)* The capability of the software product to maintain a specified level of performance in cases of software faults (defects) or of infringement of its specified interface.

Fault Tree Analysis (FTA) [ASTQB](*Synonyms: Software Fault Tree Analysis)* A technique used to analyze the causes of faults (defects). The technique visually models how logical relationships between failures, human errors, and external events can combine to cause specific faults to disclose.

Feasible path [ASTQB] A path for which a set of input values and preconditions exists which causes it to be executed.

Feature [ASTQB](*Synonyms: software feature)* An attribute of a component or system specified or implied by requirements documentation (for example reliability, usability or design constraints).

Feature-driven development [ASTQB]*(See Also: Agile software development)* An iterative and incremental software development process driven from a client-valued functionality (feature) perspective. Feature-driven development is mostly used in Agile software development.

Finished device [FDA 820.3] Any device or accessory to any device that is suitable for use or capable of functioning, whether or not it is packaged, labeled, or sterilized.

Finite state machine [ASTQB] A computational model consisting of a finite number of states and transitions between those states, possibly with accompanying actions.

Formal review [ASTQB] A review characterized by documented procedures and requirements, e.g., inspection.

Frozen test basis [ASTQB]*(See Also: baseline)* A test basis document that can only be amended by a formal change control process.

Function Point Analysis (FPA) [ASTQB] Method aiming to measure the size of the functionality of an information system. The measurement is independent of the technology. This measurement may be used as a basis for the measurement of productivity, the estimation of the needed resources, and project control.

Functional integration [ASTQB]*(See Also: integration testing)* An integration approach that combines the components or systems for the purpose of getting a basic functionality working early.

Functional requirement [ASTQB] A requirement that specifies a function that a component or system must perform.

Functional test design technique [ASTQB](*See Also: black-box test design technique)* Procedure to derive and/or select test cases based on an analysis of the specification of the functionality of a component or system without reference to its internal structure.

Functional testing [ASTQB](*See Also: black-box testing)* Testing based on an analysis of the specification of the functionality of a component or system.

Functionality [ASTQB] The capability of the software product to provide functions which meet stated and implied needs when the software is used under specified conditions.

Functionality testing [ASTQB] The process of testing to determine the functionality of a software product.

Generic test automation architecture [ASTQB] Representation of the layers, components, and interfaces of a test automation architecture, allowing for a structured and modular approach to implement test automation.

Goal Question Metric (GQM) [ASTQB] An approach to software measurement using a three-level model conceptual level (goal), operational level (question) and quantitative level (metric).

GUI [ASTQB] Acronym for Graphical User Interface.

GUI testing [ASTQB] Testing performed by interacting with the software under test via the graphical user interface.

Hardware-software integration testing [ASTQB](*See Also: integration testing)* Testing performed to expose defects in the interfaces and interaction between hardware and software components.

Hazard analysis [ASTQB](*See Also: risk analysis)* A technique used to characterize the elements of risk. The result of a hazard analysis will drive the methods used for development and testing of a system.

Heuristic evaluation [ASTQB] A usability review technique that targets usability problems in the user interface or user interface design. With this technique, the reviewers examine the interface and judge its compliance with recognized usability principles (the "heuristics").

High-level test case [ASTQB]*(See Also: low-level test case)(Synonyms: abstract test case, logical test case)* A test case without concrete (implementation level) values for input data and expected results. Logical operators are used: instances of the actual values are not yet defined and/or available.

Horizontal traceability [ASTQB] The tracing of requirements for a test level through the layers of test documentation (e.g., test plan, test design specification, test case specification and test procedure specification or test script).

Impact analysis [ASTQB] The assessment of change to the layers of development documentation, test documentation and components, in order to implement a given change to specified requirements.

Incident [ASTQB]*(Ref: After IEEE 1008 Synonyms: deviation, software test incident, test incident)* Any event occurring that requires investigation.

Incident logging [ASTQB] Recording the details of any incident that occurred, e.g., during testing.

Incident management [ASTQB]*(Ref: After IEEE 1044)* The process of recognizing, investigating, taking action and disposing of incidents. It involves logging incidents, classifying them and identifying the impact.

Incident management tool [ASTQB]*(See Also: defect management tool)* A tool that facilitates the recording and status tracking of incidents. They often have workflow oriented facilities to track and control the allocation, correction and re-testing of incidents and provide reporting facilities.

Incident report [ASTQB](*Ref: After IEEE 829 Synonyms: deviation report, software test incident report, test* incident report) A document reporting on any event that occurred, e.g., during the testing, which requires investigation.

Incremental development model [ASTQB] A development lifecycle where a project is broken into a series of increments, each of which delivers a portion of the functionality in the overall project requirements. The requirements are prioritized and delivered in priority order in the appropriate increment. In some (but not all) versions of this lifecycle model, each subproject follows a mini V-model with its own design, coding and testing phases.

Incremental testing [ASTQB] Testing where components or systems are integrated and tested one or some at a time, until all the components or systems are integrated and tested.

Independence of testing [ASTQB](*Ref: After DO-178b)* Separation of responsibilities, which encourages the accomplishment of objective testing.

Indicator [ASTQB](*Ref: ISO 14598)* A measure that can be used to estimate or predict another measure.

Infeasible path [ASTQB] A path that cannot be exercised by any set of possible input values.

Informal review [ASTQB] A review not based on a formal (documented) procedure.

Input [ASTQB] A variable (whether stored within a component or outside) that is read by a component.

Input domain [ASTQB](*See Also: domain)* The set from which valid input values can be selected.

Input value [ASTQB](*See Also: input)* An instance of an input.

Insourced testing [ASTQB] Testing performed by people who are co-located with the project team but are not fellow employees.

Inspection [ASTQB](*Ref: After IEEE 610, IEEE 1028)(See Also: peer review)* A type of peer review that relies on visual examination of documents to detect defects, e.g., violations of development standards and non-conformance to higher level documentation. The most formal review technique and therefore always based on a documented procedure.

Installability [ASTQB](*Ref: ISO 9126)(See Also: portability)* The capability of the software product to be installed in a specified environment.

Installability testing [ASTQB](*See Also: portability testing)* Testing the installability of a software product.

Installation guide [ASTQB] Supplied instructions on any suitable media, which guides the installer through the installation process. This may be a manual guide, step-by-step procedure, installation wizard, or any other similar process description.

Installation qualification (IQ) (FDA) Establishing confidence that process equipment and ancillary systems are compliant with appropriate codes and approved design intentions, and that manufacturer's recommendations are suitably considered.

installation wizard [ASTQB] Supplied software on any suitable media, which leads the installer through the installation process. It normally runs the installation process, provides feedback on installation results, and prompts for options.

Instrumentation [ASTQB] The insertion of additional code into the program in order to collect information about program behavior during execution, e.g., for measuring code coverage.

Instrumenter [ASTQB](*Synonyms: program instrumenter)* A software tool used to carry out instrumentation.

Intake test [ASTQB](*See Also: smoke test Synonyms: pretest)* A special instance of a smoke test to decide if the component or system is ready for detailed and further testing. An intake test is typically carried out at the start of the test execution phase.

Integration [ASTQB] The process of combining components or systems into larger assemblies.

Integration testing [ASTQB](*See Also: component integration testing, system integration testing*) Testing performed to expose defects in the interfaces and in the interactions between integrated components or systems.

Interface testing [ASTQB] An integration test type that is concerned with testing the interfaces between components or systems.

Interoperability [ASTQB](*Ref: After ISO 9126)(See Also: functionality*) The capability of the software product to interact with one or more specified components or systems.

Interoperability testing [ASTQB](*See Also: functionality testing)(Synonyms: compatibility testing*) Testing to determine the interoperability of a software product.

Invalid testing [ASTQB](*See Also: error tolerance, negative testing*) Testing using input values that should be rejected by the component or system.

Isolation testing [ASTQB] Testing of individual components in isolation from surrounding components, with surrounding components being simulated by stubs and drivers, if needed.

Iterative development model [ASTQB] A development lifecycle where a project is broken into a usually large number of iterations. An iteration is a complete development loop resulting in a release (internal or external) of an executable product, a subset of the final product under development, which grows from iteration to iteration to become the final product.

Keyword-driven testing [ASTQB](*See Also: data-driven testing)(Synonyms: action word-driven testing)* A scripting technique that uses data files to contain not only test data and expected results, but also keywords related to the application being tested. The keywords are interpreted by special supporting scripts that are called by the control script for the test.

Lead assessor [ASTQB] The person who leads an assessment. In some cases, for instance CMMI and TMMi when formal assessments are conducted, the lead assessor must be accredited and formally trained.

Learnability [ASTQB](*Ref: ISO 9126)(See Also: usability)* The capability of the software product to enable the user to learn its application.

Level of intrusion [ASTQB] The level to which a test object is modified by adjusting it for testability.

Level test plan [ASTQB](*See Also: test plan)* A test plan that typically addresses one test level.

Lifecycle model [ASTQB](*Ref: CMMI See Also: software lifecycle)* A partitioning of the life of a product or project into phases.

Linear scripting [ASTQB] A simple scripting technique without any control structure in the test scripts.

Load profile [ASTQB](*See Also: operational profile)* A specification of the activity which a component or system being tested may experience in production. A load profile consists of a designated number of virtual users who process a defined set of transactions in a specified time period and according to a predefined operational profile.

Load testing [ASTQB](*See Also: performance testing, stress testing)* A type of performance testing conducted to evaluate the behavior of a component or system with increasing load, e.g., numbers of parallel users and/or numbers of transactions, to determine what load can be handled by the component or system.

Load testing tool [ASTQB](*See Also: performance testing tool)* A tool to support load testing whereby it can simulate increasing load, e.g., numbers of concurrent users and/or transactions within a specified time-period.

Lot [FDA 820.3](*See Also: batch)* one or more components or finished devices that consist of a single type, model, class, size, composition, or software version that are manufactured under essentially the same conditions and that are intended to have uniform characteristics and quality within specified limits.

Low-level test case [ASTQB](*See Also: high-level test case)(Synonyms: concrete test case)* A test case with concrete (implementation level) values for input data and expected results. Logical operators from high-level test cases are replaced by actual values that correspond to the objectives of the logical operators.

Maintainability [ASTQB](*Ref: ISO 9126)* The ease with which a software product can be modified to correct defects, modified to meet new requirements, modified to make future maintenance easier, or adapted to a changed environment.

Maintainability testing [ASTQB](*Synonyms: serviceability testing)* Testing to determine the maintainability of a software product.

Maintenance [ASTQB](*Ref: IEEE 1219)* Modification of a software product after delivery to correct defects, to improve performance or other attributes, or to adapt the product to a modified environment.

Maintenance testing [ASTQB] Testing the changes to an operational system or the impact of a changed environment to an operational system.

Man-in-the-middle attack [ASTQB] The interception, mimicking and/or altering and subsequent relaying of communications (e.g., credit card transactions) by a third party such that a user remains unaware of that third party's presence.

Management review [ASTQB](*Ref: After IEEE 610, IEEE 1028)* A systematic evaluation of software acquisition, supply, development, operation, or maintenance process, performed by or on behalf of management that monitors progress, determines the status of plans and schedules, confirms requirements and their system allocation, or evaluates the effectiveness of management approaches to achieve fitness for purpose.

Manufacturing-based quality [ASTQB](*Ref: After Garvin) (See Also: product-based quality, transcendent-based quality, user-based quality, value-based quality)* A view of quality, whereby quality is measured by the degree to which a product or service conforms to its intended design and requirements. Quality arises from the process(es) used.

Master test plan [ASTQB](*See Also: test plan)* A test plan that typically addresses multiple test levels.

Maturity [ASTQB](*Ref: ISO 9126 See Also: Capability Maturity Model Integration, Test Maturity Model integration, reliability)* (1) The capability of an organization with respect to the effectiveness and efficiency of its processes and work practices. (2) The capability of the software product to avoid failure as a result of defects in the software.

Maturity level [ASTQB](*Ref: TMMi)* Degree of process improvement across a predefined set of process areas in which all goals in the set are attained.

Maturity model [ASTQB] A structured collection of elements that describe certain aspects of maturity in an organization, and aid in the definition and understanding of an organization's processes. A maturity model often provides a common language, shared vision and framework for prioritizing improvement actions.

MBT model [ASTQB] Any model used in model-based testing.

Mean time between failures (MTBF) [ASTQB](*See Also: reliability growth model)* The arithmetic mean (average) time between failures of a system. The MTBF is typically part of a reliability growth model that assumes the failed system is immediately repaired, as a part of a defect fixing process.

Mean time to repair (MTTR) [ASTQB] The arithmetic mean (average) time a system will take to recover from any failure. This typically includes testing to insure that the defect has been resolved.

Measure [ASTQB](*Ref: ISO 14598)* The number or category assigned to an attribute of an entity by making a measurement.

Measurement [ASTQB](*Ref: ISO 14598)* The process of assigning a number or category to an entity to describe an attribute of that entity.

Measurement scale [ASTQB](*Ref: ISO 14598)* A scale that constrains the type of data analysis that can be performed on it.

Memory leak [ASTQB] A memory access failure due to a defect in a program's dynamic store allocation logic that causes it to fail to release memory after it has finished using it, eventually causing the program and/or other concurrent processes to fail due to lack of memory.

Methodical test strategy [ASTQB] A test strategy whereby the test team uses a pre-determined set of test conditions such as a quality standard, a checklist, or a collection of generalized, logical test conditions which may relate to a particular domain, application or type of testing.

Methodical testing [ASTQB] Testing based on a standard set of tests, e.g., a checklist, a quality standard, or a set of generalized test cases.

Metric [ASTQB]*(Ref: ISO 1459)* A measurement scale and the method used for measurement.

Milestone [ASTQB] A point in time in a project at which defined (intermediate) deliverables and results should be ready.

Mind map [ASTQB] A diagram used to represent words, ideas, tasks, or other items linked to and arranged around a central keyword or idea. Mind maps are used to generate, visualize, structure, and classify ideas, and as an aid in study, organization, problem solving, decision making, and writing.

Model coverage [ASTQB] The degree, expressed as a percentage, to which model elements are planned to be or have been exercised by a test suite.

Model-based test strategy [ASTQB] A test strategy whereby the test team derives testware from models.

Model-based testing (MBT) [ASTQB] Testing based on or involving models.

Modeling tool [ASTQB]*(Ref: Graham)* A tool that supports the creation, amendment and verification of models of the software or system.

Moderator [ASTQB]*(Synonyms: inspection leader)* The leader and main person responsible for an inspection or other review process.

Modified condition / Decision coverage (MC/DC) [ASTQB]*(Synonyms: condition determination coverage, modified multiple condition coverage)* The percentage of all single condition outcomes that independently affect a decision outcome that have been exercised by a test case suite. 100% modified condition decision coverage implies 100% decision condition coverage.

Modified condition / Decision testing
[ASTQB](*Synonyms: condition determination testing, modified multiple condition testing)* A white-box test design technique in which test cases are designed to execute single condition outcomes that independently affect a decision outcome.

Monitoring tool [ASTQB](*Ref: After IEEE 610)* A software tool or hardware device that runs concurrently with the component or system under test and supervises, records and/or analyses the behavior of the component or system.

Monkey testing [ASTQB] Testing by means of a random selection from a large range of inputs and by randomly pushing buttons, ignorant of how the product is being used.

Multiple condition coverage [ASTQB](*Synonyms: branch condition combination coverage, condition combination coverage)* The percentage of combinations of all single condition outcomes within one statement that have been exercised by a test suite. 100% multiple condition coverage implies 100% modified condition decision coverage.

Multiple condition testing [ASTQB](*Synonyms: branch condition combination testing, condition combination testing)* A white-box test design technique in which test cases are designed to execute combinations of single condition outcomes (within one statement).

Mutation analysis [ASTQB] A method to determine test suite thoroughness by measuring the extent to which a test suite can discriminate the program from slight variants (mutants) of the program.

Mutation testing [ASTQB](*Synonyms: back-to-back testing)* Testing in which two or more variants of a component or system are executed with the same inputs, the outputs compared, and analyzed in cases of discrepancies.

Myers-Briggs Type Indicator (MBTI) [ASTQB] An indicator of psychological preference representing the different personalities and communication styles of people.

N-switch coverage [ASTQB](*Ref: Chow Synonyms: Chow's coverage metrics*) The percentage of sequences of N+1 transitions that have been exercised by a test suite.

N-switch testing [ASTQB](*Ref: Chow See Also: state transition testing*) A form of state transition testing in which test cases are designed to execute all valid sequences of N+1 transitions.

N-wise testing [ASTQB](*See Also: combinatorial testing, orthogonal array testing, pairwise testing*) A black-box test design technique in which test cases are designed to execute all possible discrete combinations of any set of n input parameters.

Negative testing [ASTQB](*Ref: After Beizer. Synonyms: dirty testing*) Tests aimed at showing that a component or system does not work. Negative testing is related to the tester's attitude rather than a specific test approach or test design technique, e.g., testing with invalid input values or exceptions.

Neighborhood integration testing [ASTQB] A form of integration testing where all of the nodes that connect to a given node are the basis for the integration testing.

Non-conformity [ASTQB](*Ref: ISO 9000*) Non-fulfillment of a specified requirement.

Non-functional requirement [ASTQB] A requirement that does not relate to functionality, but to attributes such as reliability, efficiency, usability, maintainability and portability.

Non-functional test design technique [ASTQB](*See Also: black-box test design technique*) Procedure to derive and/or select test cases for non-functional testing based on an analysis of the specification of a component or system without reference to its internal structure.

Non-functional testing [ASTQB] Testing the attributes of a component or system that do not relate to functionality, e.g., reliability, efficiency, usability, maintainability and portability.

Offline MBT [ASTQB] Model-based testing approach whereby test cases are generated into a repository for future execution.

Online MBT [ASTQB]*(Synonyms: on-the-fly MBT)* Model-based testing approach whereby test cases are generated and executed simultaneously.

Open source tool [ASTQB] A software tool that is available to all potential users in source code form, usually via the internet. +Its users are permitted, usually under license, to study, change, improve and, at times, to+ distribute the software.

Operability [ASTQB]*(Ref: ISO 9126)(See Also: usability)* The capability of the software product to enable the user to operate and control it.

Operational acceptance testing [ASTQB]*(See Also: operational testing)(Synonyms: production acceptance testing)* Operational testing in the acceptance test phase, typically performed in a (simulated) operational environment by operations and/or systems administration staff focusing on operational aspects, e.g., recoverability, resource-behavior, installability and technical compliance.

Operational environment [ASTQB] Hardware and software products installed at users' or customers' sites where the component or system under test will be used. The software may include operating systems, database management systems, and other applications.

Operational profile [ASTQB] The representation of a distinct set of tasks performed by the component or system, possibly based on user behavior when interacting with the component or system, and their probabilities of occurrence. A task is logical rather that physical and can be executed over several machines or be executed in non-contiguous time segments.

Operational profile testing [ASTQB](*Ref: Musa)* Statistical testing using a model of system operations (short duration tasks) and their probability of typical use.

Operational profiling [ASTQB](*See Also: operational profile)* The process of developing and implementing an operational profile.

Operational Qualification. (FDA) Establishing confidence that process equipment and sub-systems are capable of consistently operating within established limits and tolerances.

Operational testing [ASTQB](*Ref: IEEE 610)* Testing conducted to evaluate a component or system in its operational environment.

Orthogonal array [ASTQB] A 2-dimensional array constructed with special mathematical properties, such that choosing any two columns in the array provides every pair combination of each number in the array.

Orthogonal array testing [ASTQB](*See Also: combinatorial testing, n-wise testing, pairwise testing)* A systematic way of testing all-pair combinations of variables using orthogonal arrays. It significantly reduces the number of all combinations of variables to test all pair combinations.

Output [ASTQB] A variable (whether stored within a component or outside) that is written by a component.

Output domain [ASTQB](*See Also: domain)* The set from which valid output values can be selected.

Output value [ASTQB]*(See Also: output)* An instance of an output.

Outsourced testing [ASTQB] Testing performed by people who are not co-located with the project team and are not fellow employees.

Pair programming [ASTQB] A software development approach whereby lines of code (production and/or test) of a component are written by two programmers sitting at a single computer. This implicitly means ongoing realtime code reviews are performed.

Pair testing [ASTQB] Two persons, e.g., two testers, a developer and a tester, or an end-user and a tester, working together to find defects. Typically, they share one computer and trade control of it while testing.

Pairwise integration testing [ASTQB] A form of integration testing that targets pairs of components that work together, as shown in a call graph.

Pairwise testing [ASTQB]*(See Also: combinatorial testing, n-wise testing, orthogonal array testing)* A black-box test design technique in which test cases are designed to execute all possible discrete combinations of each pair of input parameters.

Pareto analysis [ASTQB] A statistical technique in decision making that is used for selection of a limited number of factors that produce significant overall effect. In terms of quality improvement, a large majority of problems (80%) are produced by a few key causes (20%).

Pass [ASTQB]*(Synonyms: test pass)* A test is deemed to pass if its actual result matches its expected result.

Pass/fail criteria [ASTQB]*(Ref: IEEE 829)* Decision rules used to determine whether a test item (function) or feature has passed or failed a test.

Path [ASTQB]*(Synonyms: control flow path)* A sequence of events, e.g., executable statements, of a component or system from an entry point to an exit point.

Path coverage [ASTQB] The percentage of paths that have been exercised by a test suite. 100% path coverage implies 100% LCSAJ coverage.

Path sensitizing [ASTQB] Choosing a set of input values to force the execution of a given path.

Path testing [ASTQB] A white-box test design technique in which test cases are designed to execute paths.

Peer review [ASTQB] A review of a software work product by colleagues of the producer of the product for the purpose of identifying defects and improvements. Examples are inspection, technical review and walkthrough.

Performance [ASTQB]*(Ref: After IEEE 610)(See Also: efficiency)(Synonyms: time behavior)* The degree to which a system or component accomplishes its designated functions within given constraints regarding processing time and throughput rate.

Performance indicator [ASTQB]*(Ref: CMMI Synonyms: key performance indicator)* A high-level metric of effectiveness and/or efficiency used to guide and control progressive development, e.g., lead-time slip for software development.

Performance profiling [ASTQB] The task of analyzing, e.g., identifying performance bottlenecks based on generated metrics, and tuning the performance of a software component or system using tools.

Performance testing [ASTQB]*(See Also: efficiency testing)* Testing to determine the performance of a software product.

Performance testing tool [ASTQB] A tool to support performance testing that usually has two main facilities: load generation and test transaction measurement. Load generation can simulate either

multiple users or high volumes of input data. During execution, response time measurements are taken from selected transactions and these are logged.
Performance testing tools normally provide reports based on test logs and graphs of load against response times.

Phase containment [ASTQB] The percentage of defects that are removed in the same phase of the software lifecycle in which they were introduced.

Phase test plan [ASTQB](*See Also: test plan*) A test plan that typically addresses one test phase.

Planning poker [ASTQB](*See Also: Agile software development, Wideband Delphi*) A consensus-based estimation technique, mostly used to estimate effort or relative size of user stories in Agile software development. It is a variation of the Wideband Delphi method using a deck of cards with values representing the units in which the team estimates.

Pointer [ASTQB](*Ref: IEEE 610*) A data item that specifies the location of another data item. For example, a data item that specifies the address of the next employee record to be processed.

Portability [ASTQB](*Ref: ISO 9126*) The ease with which the software product can be transferred from one hardware or software environment to another.

Portability testing [ASTQB](*Synonyms: configuration testing*) Testing to determine the portability of a software product.

Post-execution comparison [ASTQB] Comparison of actual and expected results, performed after the software has finished running.

Postcondition [ASTQB] Environmental and state conditions that must be fulfilled after the execution of a test or test procedure.

Precondition [ASTQB] Environmental and state conditions that must be fulfilled before the component or system can be executed with a particular test or test procedure.

Predicate [ASTQB](*See Also: decision*) A statement that can evaluate to true or false and may be used to determine the control flow of subsequent decision logic.

Priority [ASTQB] The level of (business) importance assigned to an item, e.g., defect.

Probe effect [ASTQB] The effect on the component or system by the measurement instrument when the component or system is being measured, e.g., by a performance testing tool or monitor. For example performance may be slightly worse when performance testing tools are being used.

Procedure testing [ASTQB] Testing aimed at ensuring that the component or system can operate in conjunction with new or existing users' business procedures or operational procedures.

Process [ASTQB](*Ref: ISO 12207*) A set of interrelated activities, which transform inputs into outputs.

Process assessment [ASTQB](*Ref: after ISO 15504*) A disciplined evaluation of an organization's software processes against a reference model.

Process cycle test [ASTQB](*Ref: TMap See Also: procedure testing*) A black-box test design technique in which test cases are designed to execute business procedures and processes.

Process improvement [ASTQB](*Ref: CMMI*) A program of activities designed to improve the performance and maturity of the organization's processes, and the result of such a program.

Process model [ASTQB] A framework wherein processes of the same nature are classified into a overall model, e.g., a test improvement model.

Process performance qualification. (FDA) Establishing confidence that the process is effective and reproducible.

Product performance qualification. (FDA) Establishing confidence through appropriate testing that the finished product produced by a specified process meets all release requirements for functionality and safety.

Process reference model [ASTQB] A process model providing a generic body of best practices and how to improve a process in a prescribed step-by-step manner.

Process-compliant test strategy [ASTQB] A test strategy whereby the test team follows a set of predefined processes, whereby the processes address such items as documentation, the proper identification and use of the test basis and test oracle(s), and the organization of the test team.

Process-compliant testing [ASTQB](*See Also: standard-compliant testing)* Testing that follows a set of defined processes, e.g., defined by an external party such as a standards committee.

Process-driven scripting [ASTQB] A scripting technique where scripts are structured into scenarios which represent use cases of the software under test. The scripts can be parameterized with test data.

Product risk [ASTQB](*See Also: risk)* A risk directly related to the test object.

Product-based quality [ASTQB](*Ref: After Garvin)(See Also: manufacturing-based quality, quality* attribute, *transcendent-based quality, user-based quality, value-based quality)* A view of quality, wherein quality is based on a well-defined set of quality attributes. These attributes must be measured in an objective and quantitative way. Differences in the quality of products of the same type can be traced back to the way the specific quality attributes have been implemented.

Project [ASTQB](*Ref: ISO 9000*) A project is a unique set of coordinated and controlled activities with start and finish dates undertaken to achieve an objective conforming to specific requirements, including the constraints of time, cost and resources.

Project retrospective [ASTQB] A structured way to capture lessons learned and to create specific action plans for improving on the next project or next project phase.

Project risk [ASTQB](*See Also: risk*) A risk related to management and control of the (test) project, e.g., lack of staffing, strict deadlines, changing requirements, etc.

Pseudo-random [ASTQB] A series which appears to be random but is in fact generated according to some prearranged sequence.

Qualification [ASTQB](*Ref: ISO 9000*) The process of demonstrating the ability to fulfill specified requirements. Note the term "qualified" is used to designate the corresponding status.

Quality [ASTQB](*Ref: After IEEE 610*) The degree to which a component, system or process meets specified requirements and/or user/customer needs and expectations.

Quality assurance [ASTQB](*Ref: ISO 9000*) Part of quality management focused on providing confidence that quality requirements will be fulfilled.

Quality attribute [ASTQB](*Ref: IEEE 610)(Synonyms: quality characteristic, software product*) characteristic, software quality characteristic A feature or characteristic that affects an item's quality.

Quality control [ASTQB](*Ref: after ISO 8402*) The operational techniques and activities, part of quality management, that are focused on fulfilling quality requirements.

Quality function deployment (QFD) [ASTQB](*Ref: Akao*) A method to transform user demands into design quality, to deploy the functions forming quality, and to deploy methods for achieving the design quality into subsystems and component parts, and ultimately to specific elements of the manufacturing process.

Quality gate [ASTQB] A special milestone in a project. Quality gates are located between those phases of a project strongly depending on the outcome of a previous phase. A quality gate includes a formal check of the documents of the previous phase.

Quality management [ASTQB](*Ref: ISO 9000*) Coordinated activities to direct and control an organization with regard to quality. Direction and control with regard to quality generally includes the establishment of the quality policy and quality objectives, quality planning, quality control, quality assurance and quality improvement.

Quality risk [ASTQB](*See Also: quality attribute, product risk*) A product risk related to a quality attribute.

RACI matrix [ASTQB] A matrix describing the participation by various roles in completing tasks or deliverables for a project or process. It is especially useful in clarifying roles and responsibilities. RACI is an acronym derived from the four key responsibilities most typically used: Responsible, Accountable, Consulted, and Informed.

Random testing [ASTQB] A black-box test design technique where test cases are selected, possibly using a pseudo-random generation algorithm, to match an operational profile. This technique can be used for testing nonfunctional attributes such as reliability and performance.

Reactive test strategy [ASTQB] A test strategy whereby the test team waits to design and implement tests until the software is received, reacting to the actual system under test.

Reactive testing [ASTQB] Testing that dynamically responds to the actual system under test and test results being obtained. Typically reactive testing has a reduced planning cycle and the design and implementation test phases are not carried out until the test object is received.

Recoverability [ASTQB](*Ref: ISO 9126)(See Also: reliability)* The capability of the software product to re-establish a specified level of performance and recover the data directly affected in case of failure.

Recoverability testing [ASTQB](*See Also: reliability testing)(Synonyms: recovery testing)* Testing to determine the recoverability of a software product.

Regression testing [ASTQB] Testing of a previously tested program following modification to ensure that defects have not been introduced or uncovered in unchanged areas of the software, as a result of the changes made. It is performed when the software or its environment is changed.

Regression-averse test strategy [ASTQB] A test strategy whereby the test team applies various techniques to manage the risk of regression such as functional and/or non-functional regression test automation at one or more levels.

Regression-averse testing [ASTQB] Testing using various techniques to manage the risk of regression, e.g., by designing re-usable testware and by extensive automation of testing at one or more test levels.

Release note [ASTQB](*Ref: After IEEE 829)(Synonyms: item transmittal report, test item transmittal report)* A document identifying test items, their configuration, current status and other delivery information delivered by development to testing, and possibly other stakeholders, at the start of a test execution phase.

Reliability [ASTQB]*(Ref: ISO 9126)* The ability of the software product to perform its required functions under stated conditions for a specified period of time, or for a specified number of operations.

Reliability growth model [ASTQB] A model that shows the growth in reliability over time during continuous testing of a component or system as a result of the removal of defects that result in reliability failures.

Reliability testing [ASTQB] Testing to determine the reliability of a software product.

Replaceability [ASTQB]*(Ref: ISO 9126)(See Also: portability)* The capability of the software product to be used in place of another specified software product for the same purpose in the same environment.

Requirement [ASTQB]*(Ref: After IEEE 610)* A condition or capability needed by a user to solve a problem or achieve an objective that must be met or possessed by a system or system component to satisfy a contract, standard, specification, or other formally imposed document.

Requirements management tool [ASTQB] A tool that supports the recording of requirements, requirements attributes (e.g., priority, knowledge responsible) and annotation, and facilitates traceability through layers of requirements and requirements change management. Some requirements management tools also provide facilities for static analysis, such as consistency checking and violations to predefined requirements rules.

Requirements phase [ASTQB]*(Ref: IEEE 610)* The period of time in the software lifecycle during which the requirements for a software product are defined and documented.

Requirements-based testing [ASTQB] An approach to testing in which test cases are designed based on test objectives and test conditions derived from requirements, e.g., tests that exercise specific functions or probe nonfunctional attributes such as reliability or usability.

Resource utilization [ASTQB]*(Ref: After ISO 9126)(See Also: efficiency Synonyms: storage)* The capability of the software product to use appropriate amounts and types of resources, for example the amounts of main and secondary memory used by the program and the sizes of required temporary or overflow files, when the software performs its function under stated conditions.

Resource utilization testing [ASTQB]*(See Also: efficiency testing)(Synonyms: storage testing)* The process of testing to determine the resource-utilization of a software product.

Result [ASTQB]*(See Also: actual result, expected result)(Synonyms: outcome, test outcome, test result)* The consequence/outcome of the execution of a test. It includes outputs to screens, changes to data, reports, and communication messages sent out.

Resumption criteria [ASTQB] The criteria used to restart all or a portion of the testing activities that were suspended previously.

Resumption requirements [ASTQB]*(Ref: After IEEE 829)* The defined set of testing activities that must be repeated when testing is re-started after a suspension.

Retrospective meeting [ASTQB]*(Synonyms: post-project meeting)* A meeting at the end of a project during which the project team members evaluate the project and learn lessons that can be applied to the next project.

Review [ASTQB](*Ref: After IEEE 1028)* An evaluation of a product or project status to ascertain discrepancies from planned results and to recommend improvements. Examples include management review, informal review, technical review, inspection, and walkthrough.

Review plan [ASTQB] A document describing the approach, resources and schedule of intended review activities. It identifies, amongst others: documents and code to be reviewed, review types to be used, participants, as well as entry and exit criteria to be applied in case of formal reviews, and the rationale for their choice. It is a record of the review planning process.

Review tool [ASTQB] A tool that provides support to the review process. Typical features include review planning and tracking support, communication support, collaborative reviews and a repository for collecting and reporting of metrics.

Reviewer [ASTQB](*Synonyms: checker, inspector)* The person involved in the review that identifies and describes anomalies in the product or project under review. Reviewers can be chosen to represent different viewpoints and roles in the review process.

Risk [ASTQB] A factor that could result in future negative consequences.

Risk analysis [ASTQB] The process of assessing identified project or product risks to determine their level of risk, typically by estimating their impact and probability of occurrence (likelihood).

Risk assessment [ASTQB](*See Also: product risk, project risk, risk, risk impact, risk level, risk likelihood)* The process of identifying and subsequently analyzing the identified project or product risk to determine its level of risk, typically by assigning likelihood and impact ratings.

Risk identification [ASTQB] The process of identifying risks using techniques such as brainstorming, checklists and failure history.

Risk impact [ASTQB](*Synonyms: impact*) The damage that will be caused if the risk becomes an actual outcome or event.

Risk level The importance of a risk as defined by its characteristics impact and likelihood. The level of risk can be used to determine the intensity of testing to be performed. A risk level can be expressed either qualitatively (e.g., high, medium, low) or quantitatively.

Risk likelihood [ASTQB](*Synonyms: likelihood*) The estimated probability that a risk will become an actual outcome or event.

Risk management [ASTQB] Systematic application of procedures and practices to the tasks of identifying, analyzing, prioritizing, and controlling risk.

Risk mitigation [ASTQB](*Synonyms: risk control*) The process through which decisions are reached and protective measures are implemented for reducing risks to, or maintaining risks within, specified levels.

Risk type [ASTQB](*Synonyms: risk category*) A set of risks grouped by one or more common factors such as a quality attribute, cause, location, or potential effect of risk. A specific set of product risk types is related to the type of testing that can mitigate (control) that risk type. For example, the risk of user interactions being misunderstood can be mitigated by usability testing.

Risk-based testing [ASTQB] An approach to testing to reduce the level of product risks and inform stakeholders of their status, starting in the initial stages of a project. It involves the identification of product risks and the use of risk levels to guide the test process.

Robustness [ASTQB]*(Ref: IEEE 610)(See Also: error-tolerance, fault-tolerance)* The degree to which a component or system can function correctly in the presence of invalid inputs or stressful environmental conditions.

Robustness testing [ASTQB] Testing to determine the robustness of the software product.

Root cause [ASTQB]*(Ref: CMMI)* A source of a defect such that if it is removed, the occurrence of the defect type is decreased or removed.

Root cause analysis [ASTQB] An analysis technique aimed at identifying the root causes of defects. By directing corrective measures at root causes, it is hoped that the likelihood of defect recurrence will be minimized.

S.M.A.R.T. goal methodology (SMART) [ASTQB] A methodology whereby objectives are defined very specifically rather than generically. SMART is an acronym derived from the attributes of the objective to be defined: Specific, Measurable, Attainable, Relevant and Timely.

Safety [ASTQB]*(Ref: ISO 9126)* The capability of the software product to achieve acceptable levels of risk of harm to people, business, software, property or the environment in a specified context of use.

Safety critical system [ASTQB] A system whose failure or malfunction may result in death or serious injury to people, or loss or severe damage to equipment, or environmental harm.

Safety testing [ASTQB] Testing to determine the safety of a software product.

Scalability [ASTQB]*(Ref: After Gerrard)* The capability of the software product to be upgraded to accommodate increased loads.

Scalability testing [ASTQB] Testing to determine the scalability of the software product.

Scorecard [ASTQB](*See Also: balanced scorecard, dashboard*) A representation of summarized performance measurements representing progress towards the implementation of long-term goals. A scorecard provides static measurements of performance over or at the end of a defined interval.

Scribe [ASTQB](*Synonyms: recorder*) The person who records each defect mentioned and any suggestions for process improvement during a review meeting, on a logging form. The scribe should ensure that the logging form is readable and understandable.

Scripted testing [ASTQB] Test execution carried out by following a previously documented sequence of tests.

Scripting language [ASTQB] A programming language in which executable test scripts are written, used by a test execution tool (e.g., a capture/playback tool).

Scrum [ASTQB](*See Also: Agile software development*) An iterative incremental framework for managing projects commonly used with Agile software development.

Security [ASTQB](*Ref: ISO 9126 See Also: functionality*) Attributes of software products that bear on its ability to prevent unauthorized access, whether accidental or deliberate, to programs and data.

Security testing [ASTQB](*See Also: functionality testing*) Testing to determine the security of the software product.

Security testing tool [ASTQB] A tool that provides support for testing security characteristics and vulnerabilities.

Security tool [ASTQB] A tool that supports operational security.

Session-based test management [ASTQB] A method for measuring and managing session-based testing, e.g., exploratory testing.

Session-based testing [ASTQB] An approach to testing in which test activities are planned as uninterrupted sessions of test design and execution, often used in conjunction with exploratory testing.

Severity [ASTQB]*(Ref: After IEEE 610)* The degree of impact that a defect has on the development or operation of a component or system.

Short-circuiting [ASTQB] A programming language/interpreter technique for evaluating compound conditions in which a condition on one side of a logical operator may not be evaluated if the condition on the other side is sufficient to determine the final outcome.

Simulation [ASTQB]*(Ref: ISO 2382/1)* The representation of selected behavioral characteristics of one physical or abstract system by another system.

Simulator [ASTQB]*(Ref: After IEEE 610, DO178b)(See Also: emulator)* A device, computer program or system used during testing, which behaves or operates like a given system when provided with a set of controlled inputs.

Site acceptance testing [ASTQB] Acceptance testing by users/customers at their site, to determine whether or not a component or system satisfies the user/customer needs and fits within the business processes, normally including hardware as well as software.

Smoke test [ASTQB]*(See Also: build, verification test, intake test)(Synonyms: confidence test, sanity test)* A subset of all defined/planned test cases that cover the main functionality of a component or system, to ascertaining that the most crucial functions of a program work, but not bothering with finer details.

Software [ASTQB]*(Ref: IEEE 610)* Computer programs, procedures, and possibly associated documentation and data pertaining to the operation of a computer system.

Software integrity level [ASTQB] The degree to which software complies or must comply with a set of stakeholder-selected software and/or software-based system characteristics (e.g., software complexity, risk assessment, safety level, security level, desired performance, reliability or cost) which are defined to reflect the importance of the software to its stakeholders.

Software lifecycle [ASTQB] The period of time that begins when a software product is conceived and ends when the software is no longer available for use. The software lifecycle typically includes a concept phase, requirements phase, design phase, implementation phase, test phase, installation and checkout phase, operation and maintenance phase, and sometimes, retirement phase. Note these phases may overlap or be performed iteratively.

Software process improvement (SPI) [ASTQB]*(Ref: After CMMI)* A program of activities designed to improve the performance and maturity of the organization's software processes and the results of such a program.

software quality [ASTQB]*(Ref: After ISO 9126)(See Also: quality)* The totality of functionality and features of a software product that bear on its ability to satisfy stated or implied needs.

Software Usability Measurement Inventory (SUMI) [ASTQB]*(Ref: Kirakowski93)* A questionnaire-based usability test technique for measuring software quality from the end user's point of view.

Specification [ASTQB]*(Ref: After IEEE 610)* A document that specifies, ideally in a complete, precise and verifiable manner, the requirements, design, behavior, or other characteristics of a component or system, and, often, the procedures for determining whether these provisions have been satisfied.

Specified input [ASTQB] An input for which the specification predicts a result.

Stability [ASTQB]*(Ref: ISO 9126)(See Also: maintainability)* The capability of the software product to avoid unexpected effects from modifications in the software.

Staged representation [ASTQB]*(See Also: CMMI)* A model structure wherein attaining the goals of a set of process areas establishes a maturity level; each level builds a foundation for subsequent levels.

Standard [ASTQB]*(Ref: After CMMI)* Formal, possibly mandatory, set of requirements developed and used to prescribe consistent approaches to the way of working or to provide guidelines (e.g., ISO/IEC standards, IEEE standards, and organizational standards).

Standard-compliant test strategy [ASTQB] A test strategy whereby the test team follows a standard. Standards followed may be valid e.g., for a country (legislation standards), a business domain (domain standards), or internally (organizational standards).

Standard-compliant testing [ASTQB]*(See Also: process-compliant testing)* Testing that complies to a set of requirements defined by a standard, e.g., an industry testing standard or a standard for testing safety-critical systems.

State diagram [ASTQB]*(Ref: IEEE 610)* A diagram that depicts the states that a component or system can assume, and shows the events or circumstances that cause and/or result from a change from one state to another.

State table [ASTQB] A grid showing the resulting transitions for each state combined with each possible event, showing both valid and invalid transitions.

State transition [ASTQB] A transition between two states of a component or system.

State transition testing [ASTQB](*See Also: N-switch testing)(Synonyms: finite state testing)* A black-box test design technique in which test cases are designed to execute valid and invalid state transitions.

Statement [ASTQB](*Synonyms: source statement)* An entity in a programming language, which is typically the smallest indivisible unit of execution.

Statement coverage [ASTQB] The percentage of executable statements that have been exercised by a test suite.

Statement testing [ASTQB] A white-box test design technique in which test cases are designed to execute statements.

Static analysis [ASTQB] Analysis of software development artifacts, e.g., requirements or code, carried out without execution of these software development artifacts. Static analysis is usually carried out by means of a supporting tool.

Static analyzer [ASTQB](*Synonyms: analyzer, static analysis tool)* A tool that carries out static analysis.

static code analysis [ASTQB] Analysis of source code carried out without execution of that software.

Static testing [ASTQB] Testing of a software development artifact, e.g., requirements, design or code, without execution of these artifacts, e.g., reviews or static analysis.

Statistical testing [ASTQB](*See Also: operational profile testing)* A test design technique in which a model of the statistical distribution of the input is used to construct representative test cases.

Status accounting [ASTQB](*Ref: IEEE 610)* An element of configuration management consisting of the recording and reporting of information needed to manage a configuration effectively. This information includes a listing of the approved configuration identification, the status of proposed changes to the configuration, and the implementation status of the approved changes.

Stress testing [ASTQB](*Ref: After IEEE 610 See Also: performance testing, load testing*) A type of performance testing conducted to evaluate a system or component at or beyond the limits of its anticipated or specified workloads, or with reduced availability of resources such as access to memory or servers.

Stress testing tool [ASTQB] A tool that supports stress testing.

Structural coverage [ASTQB] Coverage measures based on the internal structure of a component or system.

Structured scripting [ASTQB] A scripting technique that builds and utilizes a library of reusable (parts of) scripts.

Stub [ASTQB](*Ref: After IEEE 610*) A skeletal or special-purpose implementation of a software component, used to develop or test a component that calls or is otherwise dependent on it. It replaces a called component.

Subpath [ASTQB] A sequence of executable statements within a component.

Suitability [ASTQB](*Ref: ISO 9126)(See Also: functionality*) The capability of the software product to provide an appropriate set of functions for specified tasks and user objectives.

Suitability testing [ASTQB] Testing to determine the suitability of a software product.

Suspension criteria [ASTQB](*Ref: After IEEE 829*) The criteria used to (temporarily) stop all or a portion of the testing activities on the test items.

Syntax testing [ASTQB] A black-box test design technique in which test cases are designed based upon the definition of the input domain and/or output domain.

System [ASTQB](*Ref: IEEE 610)* A collection of components organized to accomplish a specific function or set of functions.

System integration testing [ASTQB] Testing the integration of systems and packages; testing interfaces to external organizations (e.g., Electronic Data Interchange, Internet).

System of systems [ASTQB] Multiple heterogeneous, distributed systems that are embedded in networks at multiple levels and in multiple interconnected domains, addressing large-scale inter-disciplinary common problems and purposes, usually without a common management structure.

System testing [ASTQB](*Ref: Hetzel)* Testing an integrated system to verify that it meets specified requirements.

System under test (SUT) [ASTQB] See test object.

Systematic Test and Evaluation Process (STEP) [ASTQB](*See Also: content-based model)* A structured testing methodology, also used as a content-based model for improving the testing process. Systematic Test and Evaluation Process (STEP) does not require that improvements occur in a specific order.

Technical review [ASTQB](*Ref: Gilb and Graham, IEEE 1028)(See Also: peer review)* A peer group discussion activity that focuses on achieving consensus on the technical approach to be taken.

Test [ASTQB](*Ref: IEEE 829)* A set of one or more test cases.

Test adaptation layer [ASTQB] The layer in a test automation architecture which provides the necessary code to adapt test scripts on an abstract level to the various components, configuration or interfaces of the SUT.

Test analysis [ASTQB] The process of analyzing the test basis and defining test objectives.

Test approach [ASTQB] The implementation of the test strategy for a specific project. It typically includes the decisions made that follow based on the (test) project's goal and the risk assessment carried out, starting points regarding the test process, the test design techniques to be applied, exit criteria and test types to be performed.

Test architect [ASTQB] (1) A person who provides guidance and strategic direction for a test organization and for its relationship with other disciplines. (2) A person who defines the way testing is structured for a given system, including topics such as test tools and test data management.

Test automation [ASTQB] The use of software to perform or support test activities, e.g., test management, test design, test execution and results checking.

Test automation architecture [ASTQB] An instantiation of the generic test automation architecture to define the architecture of a test automation solution, i.e., its layers, components, services and interfaces.

Test automation engineer [ASTQB] A person who is responsible for the design, implementation and maintenance of a test automation architecture as well as the technical evolution of the resulting test automation solution.

Test automation framework [ASTQB] A tool that provides an environment for test automation. It usually includes a test harness and test libraries.

Test automation manager [ASTQB] A person who is responsible for the planning and supervision of the development and evolution of a test automation solution.

Test automation solution [ASTQB] A realization/implementation of a test automation architecture, i.e., a combination of components implementing a specific test automation assignment. The components may include off-the-shelf test tools, test automation frameworks, as well as test hardware.

Test automation strategy [ASTQB] A high-level plan to achieve long-term objectives of test automation under given boundary conditions.

Test basis [ASTQB](*Ref: After TMap)* All documents from which the requirements of a component or system can be inferred. The documentation on which the test cases are based. If a document can be amended only by way of formal amendment procedure, then the test basis is called a frozen test basis.

Test case [ASTQB](*Ref: After IEEE 610)* A set of input values, execution preconditions, expected results and execution postconditions, developed for a particular objective or test condition, such as to exercise a particular program path or to verify compliance with a specific requirement.

Test case explosion [ASTQB] The disproportionate growth of the number of test cases with growing size of the test basis, when using a certain test design technique. Test case explosion may also happen when applying the test design technique systematically for the first time.

Test case result [ASTQB] The final verdict on the execution of a test and its outcomes, such as pass, fail, or error. The result of error is used for situations where it is not clear whether the problem is in the test object.

Test case specification [ASTQB](*Ref: After IEEE 829)(See Also: test specification)* A document specifying a set of test cases (objective, inputs, test actions, expected results, and execution preconditions) for a test item.

Test charter [ASTQB](*See Also: exploratory testing)(Synonyms: charter)* A statement of test objectives, and possibly test ideas about how to test. Test charters are used in exploratory testing.

Test closure [ASTQB](*See Also: test process)* During the test closure phase of a test process data is collected from completed activities to consolidate experience, testware, facts and numbers. The test closure phase consists of finalizing and archiving the testware and evaluating the test process, including preparation of a test evaluation report.

Test comparator [ASTQB](*Synonyms: comparator)* A test tool to perform automated test comparison of actual results with expected results.

Test comparison [ASTQB] The process of identifying differences between the actual results produced by the component or system under test and the expected results for a test. Test comparison can be performed during test execution (dynamic comparison) or after test execution.

Test condition [ASTQB](*Synonyms: test requirement, test situation)* An item or event of a component or system that could be verified by one or more test cases, e.g., a function, transaction, feature, quality attribute, or structural element.

Test control [ASTQB](*See Also: test management)* A test management task that deals with developing and applying a set of corrective actions to get a test project on track when monitoring shows a deviation from what was planned.

Test cycle [ASTQB] Execution of the test process against a single identifiable release of the test object.

Test data [ASTQB] Data that exists (for example, in a database) before a test is executed, and that affects or is affected by the component or system under test.

Test data management [ASTQB] The process of analyzing test data requirements, designing test data structures, creating and maintaining test data.

Test data preparation tool [ASTQB](*Synonyms: test generator*) A type of test tool that enables data to be selected from existing databases or created, generated, manipulated and edited for use in testing.

Test definition layer [ASTQB] The layer in a generic test automation architecture which supports test implementation by supporting the definition of test suites and/or test cases, e.g., by offering templates or guidelines.

Test deliverable [ASTQB](*See Also: deliverable*) Any test (work) product that must be delivered to someone other than the test (work) product's author.

Test design [ASTQB](*See Also: test design specification*) The process of transforming general test objectives into tangible test conditions and test cases.

Test design specification [ASTQB](*Ref: After IEEE 829)(See Also: test specification*) A document specifying the test conditions (coverage items) for a test item, the detailed test approach and identifying the associated high-level test cases.

Test design technique [ASTQB](*Synonyms: test case design technique, test specification* technique*, test technique*) Procedure used to derive and/or select test cases.

Test design tool [ASTQB] A tool that supports the test design activity by generating test inputs from a specification that may be held in a CASE tool repository, e.g., requirements management tool, from specified test conditions held in the tool itself, or from code.

Test director [ASTQB](*See Also: test manager*) A senior manager who manages test managers.

Test environment [ASTQB](*Ref: After IEEE 610)(Synonyms: test bed, test rig*) An environment containing hardware, instrumentation, simulators, software tools, and other support elements needed to conduct a test.

Test estimation [ASTQB] The calculated approximation of a result related to various aspects of testing (e.g., effort spent, completion date, costs involved, number of test cases, etc.) which is usable even if input data may be incomplete, uncertain, or noisy.

Test evaluation report [ASTQB] A document produced at the end of the test process summarizing all testing activities and results. It also contains an evaluation of the test process and lessons learned.

Test execution [ASTQB] The process of running a test on the component or system under test, producing actual result(s).

Test execution automation [ASTQB] The use of software, e.g., capture/playback tools, to control the execution of tests, the comparison of actual results to expected results, the setting up of test preconditions, and other test control and reporting functions.

Test execution layer [ASTQB] The layer in a generic test automation architecture which supports the execution of test suites and/or test cases.

Test execution phase [ASTQB]*(Ref: IEEE 610)* The period of time in a software development lifecycle during which the components of a software product are executed, and the software product is evaluated to determine whether or not requirements have been satisfied.

Test execution schedule [ASTQB] A scheme for the execution of test procedures. Note: The test procedures are included in the test execution schedule in their context and in the order in which they are to be executed.

Test execution technique [ASTQB] The method used to perform the actual test execution, either manual or automated.

Test execution tool [ASTQB] A type of test tool that is able to execute other software using an automated test script, e.g., capture/playback.

Test generation layer [ASTQB] The layer in a generic test automation architecture which supports manual or automated design of test suites and/or test cases.

Test harness [ASTQB] A test environment comprised of stubs and drivers needed to execute a test.

Test hook [ASTQB] A customized software interface that enables automated testing of a test object.

Test implementation [ASTQB] The process of developing and prioritizing test procedures, creating test data and, optionally, preparing test harnesses and writing automated test scripts.

Test improvement plan [ASTQB]*(Ref: After CMMI)* A plan for achieving organizational test process improvement objectives based on a thorough understanding of the current strengths and weaknesses of the organization's test processes and test process assets.

Test infrastructure [ASTQB] The organizational artifacts needed to perform testing, consisting of test environments, test tools, office environment and procedures.

Test input [ASTQB] The data received from an external source by the test object during test execution. The external source can be hardware, software or human.

Test item [ASTQB]*(See Also: test object)* The individual element to be tested. There usually is one test object and many test items.

Test level [ASTQB]*(Ref: After TMap)(Synonyms: test stage)* A group of test activities that are organized and managed together. A test level is linked to the responsibilities in a project. Examples of test levels are component test, integration test, system test and acceptance test.

Test log [ASTQB](*Ref: IEEE 829)(Synonyms: test record, test run log)* A chronological record of relevant details about the execution of tests.

Test logging [ASTQB](*Synonyms: test recording)* The process of recording information about tests executed into a test log.

Test management [ASTQB] The planning, estimating, monitoring and control of test activities, typically carried out by a test manager.

Test management tool [ASTQB] A tool that provides support to the test management and control part of a test process. It often has several capabilities, such as testware management, scheduling of tests, the logging of results, progress tracking, incident management and test reporting.

Test manager [ASTQB](*Synonyms: test leader)* The person responsible for project management of testing activities and resources, and evaluation of a test object. The individual who directs, controls, administers, plans and regulates the evaluation of a test object.

Test mission [ASTQB](*See Also: test policy)* The purpose of testing for an organization, often documented as part of the test policy.

Test model [ASTQB] A model describing testware that is used for testing a component or a system under test.

Test monitoring [ASTQB](*See Also: test management)* A test management task that deals with the activities related to periodically checking the status of a test project. Reports are prepared that compare the actuals to that which was planned.

Test object [ASTQB](*See Also: test item Synonyms: system under test)* The component or system to be tested.

Test objective [ASTQB] A reason or purpose for designing and executing a test.

Test oracle [ASTQB]*(Ref: After Adrion)(Synonyms: oracle)* A source to determine expected results to compare with the actual result of the software under test. An oracle may be the existing system (for a benchmark), other software, a user manual, or an individual's specialized knowledge, but should not be the code.

Test performance indicator [ASTQB] A high-level metric of effectiveness and/or efficiency used to guide and control progressive test development, e.g., Defect Detection Percentage (DDP).

Test phase [ASTQB]*(Ref: After Gerrard)* A distinct set of test activities collected into a manageable phase of a project, e.g., the execution activities of a test level.

Test plan [ASTQB]*(Ref: After IEEE 829)* A document describing the scope, approach, resources and schedule of intended test activities. It identifies amongst others test items, the features to be tested, the testing tasks, who will do each task, degree of tester independence, the test environment, the test design techniques and entry and exit criteria to be used, and the rationale for their choice, and any risks requiring contingency planning. It is a record of the test planning process.

Test planning [ASTQB] The activity of establishing or updating a test plan.

Test Point Analysis (TPA) [ASTQB]*(Ref: TMap)* A formula based test estimation method based on function point analysis.

Test policy [ASTQB] A high-level document describing the principles, approach and major objectives of the organization regarding testing.

Test procedure specification [ASTQB]*(Ref: After IEEE 829)(See Also: test specification)(Synonyms: test procedure, test scenario)* A document specifying a sequence of actions for the execution of a test. Also known as test script or manual test script.

Test process [ASTQB] The fundamental test process comprises test planning and control, test analysis and design, test implementation and execution, evaluating exit criteria and reporting, and test closure activities.

Test process group (TPG) [ASTQB]*(Ref: After CMMI)* A collection of (test) specialists who facilitate the definition, maintenance, and improvement of the test processes used by an organization.

Test process improvement [ASTQB]*(Ref: After CMMI)* A program of activities designed to improve the performance and maturity of the organization's test processes and the results of such a program.

Test process improvement manifesto [ASTQB]*(Ref: Veenendaal08)* A statement that echoes the Agile manifesto, and defines values for improving the testing process. The values are: flexibility over detailed processes, best practices over templates, deployment orientation over process orientation, peer reviews over quality assurance (departments), business driven over model-driven.

Test process improver [ASTQB] A person implementing improvements in the test process based on a test improvement plan.

Test progress report [ASTQB]*(Synonyms: test report)* A document summarizing testing activities and results, produced at regular intervals, to report progress of testing activities against a baseline (such as the original test plan) and to communicate risks and alternatives requiring a decision to management.

Test reporting [ASTQB]*(See Also: test process)* Collecting and analyzing data from testing activities and subsequently consolidating the data in a report to inform stakeholders.

Test reproducibility [ASTQB] An attribute of a test indicating whether the same results are produced each time the test is executed.

Test run [ASTQB] Execution of a test on a specific version of the test object.

Test schedule [ASTQB] A list of activities, tasks or events of the test process, identifying their intended start and finish dates and/or times, and interdependencies.

Test script [ASTQB] Commonly used to refer to a test procedure specification, especially an automated one.

Test selection criteria [ASTQB] The criteria used to guide the generation of test cases or to select test cases in order to limit the size of a test.

Test session [ASTQB]*(See Also: exploratory testing)* An uninterrupted period of time spent in executing tests. In exploratory testing, each test session is focused on a charter, but testers can also explore new opportunities or issues during a session. The tester creates and executes on the fly and records their progress.

Test specification [ASTQB] A document that consists of a test design specification, test case specification and/or test procedure specification.

Test strategy [ASTQB] A high-level description of the test levels to be performed and the testing within those levels for an organization or programme (one or more projects).

Test suite [ASTQB]*(Synonyms: test case suite, test set)* A set of several test cases for a component or system under test, where the post condition of one test is often used as the precondition for the next one.

Test summary report [ASTQB]*(Ref: After IEEE 829)(Synonyms: test report)* A document summarizing testing activities and results. It also contains an evaluation of the corresponding test items against exit criteria.

Test target [ASTQB] A set of exit criteria.

Test tool [ASTQB]*(Ref: TMap)(See Also: CAST)* A software product that supports one or more test activities, such as planning and control, specification, building initial files and data, test execution and test analysis.

Test type [ASTQB]*(Ref: After TMap)* A group of test activities aimed at testing a component or system focused on a specific test objective, i.e. functional test, usability test, regression test etc. A test type may take place on one or more test levels or test phases.

Test-driven development (TDD) [ASTQB] A way of developing software where the test cases are developed, and often automated, before the software is developed to run those test cases.

Testability [ASTQB]*(Ref: ISO 9126)(See Also: maintainability)* The capability of the software product to enable modified software to be tested.

Testability review [ASTQB]*(Ref: After TMap)* A detailed check of the test basis to determine whether the test basis is at an adequate quality level to act as an input document for the test process.

Testable requirement [ASTQB]*(Ref: After IEEE 610)* A requirement that is stated in terms that permit establishment of test designs (and subsequently test cases) and execution of tests to determine whether the requirement has been met.

Tester [ASTQB] A skilled professional who is involved in the testing of a component or system.

Testing [ASTQB] The process consisting of all lifecycle activities, both static and dynamic, concerned with planning, preparation and evaluation of software products and related work products to determine that they satisfy specified requirements, to demonstrate that they are fit for purpose and to detect defects.

Testware [ASTQB](*Ref: After Fewster and Graham)* Artifacts produced during the test process required to plan, design, and execute tests, such as documentation, scripts, inputs, expected results, set-up and clear-up procedures, files, databases, environment, and any additional software or utilities used in testing.

Thread testing [ASTQB] An approach to component integration testing where the progressive integration of components follows the implementation of subsets of the requirements, as opposed to the integration of components by levels of a hierarchy.

Three-point estimation [ASTQB] A test estimation method using estimated values for the "best case", "worst case", and "most likely case" of the matter being estimated, to define the degree of certainty associated with the resultant estimate.

Top-down testing [ASTQB](*See Also: integration testing)* An incremental approach to integration testing where the component at the top of the component hierarchy is tested first, with lower level components being simulated by stubs. Tested components are then used to test lower level components. The process is repeated until the lowest level components have been tested.

Total Quality Management (TQM) [ASTQB](*Ref: After ISO 8402)* An organization-wide management approach centered on quality, based on the participation of all members of the organization and aiming at long-term success through customer satisfaction, and benefits to all members of the organization and to society. Total Quality Management consists of planning, organizing, directing, control, and assurance.

Traceability [ASTQB](*See Also: horizontal traceability, vertical traceability*) The ability to identify related items in documentation and software, such as requirements with associated tests.

Traceability matrix [ASTQB] A two-dimensional table, which correlates two entities (e.g., requirements and test cases). The table allows tracing back and forth the links of one entity to the other, thus enabling the determination of coverage achieved and the assessment of impact of proposed changes.

Transcendent-based quality [ASTQB](*Ref: After Garvin)(See Also: manufacturing-based quality, product-based quality, user-based quality, value-based quality*) A view of quality, wherein quality cannot be precisely defined, but we know it when we see it, or are aware of its absence when it is missing. Quality depends on the perception and affective feelings of an individual or group of individuals toward a product.

Understandability [ASTQB](*Ref: ISO 9126)(See Also: usability*) The capability of the software product to enable the user to understand whether the software is suitable, and how it can be used for particular tasks and conditions of use.

Unit test framework [ASTQB](*Ref: Graham*) A tool that provides an environment for unit or component testing in which a component can be tested in isolation or with suitable stubs and drivers. It also provides other support for the developer, such as debugging capabilities.

Unreachable code [ASTQB](*Synonyms: dead code*) Code that cannot be reached and therefore is impossible to execute.

Usability [ASTQB](*Ref: ISO 9126*) The capability of the software to be understood, learned, used and attractive to the user when used under specified conditions.

Usability testing [ASTQB](*Ref: After ISO 9126)* Testing to determine the extent to which the software product is understood, easy to learn, easy to operate and attractive to the users under specified conditions.

Use case [ASTQB] A sequence of transactions in a dialogue between an actor and a component or system with a tangible result, where an actor can be a user or anything that can exchange information with the system.

Use case testing [ASTQB](*Synonyms: scenario testing, user scenario testing)* A black-box test design technique in which test cases are designed to execute scenarios of use cases.

User acceptance testing [ASTQB](*See Also: acceptance testing)* Acceptance testing carried out by future users in a (simulated) operational environment focusing on user requirements and needs.

User story [ASTQB](*See Also: Agile software development, requirement)* A high-level user or business requirement commonly used in Agile software development, typically consisting of one or more sentences in the everyday or business language capturing what functionality a user needs, any non-functional criteria, and also includes acceptance criteria.

User story testing [ASTQB](*See Also: user story)* A black-box test design technique in which test cases are designed based on user stories to verify their correct implementation.

User test [ASTQB] A test whereby real-life users are involved to evaluate the usability of a component or system.

User-based quality [ASTQB](*Ref: after Garvin See Also: manufacturing-based quality, product-based quality, transcendent-based quality, valuebased quality)* A view of quality, wherein quality is the capacity to satisfy needs, wants and desires of the user(s). A product or service that does not fulfill user needs is unlikely to find any users. This is a context dependent, contingent approach to quality since different business characteristics require different qualities of a product.

V-model [ASTQB] A framework to describe the software development lifecycle activities from requirements specification to maintenance. The V-model illustrates how testing activities can be integrated into each phase of the software development lifecycle.

V&V verification and validation.

VV&T validation, verification, and testing.

Validation [ASTQB](*Ref: ISO 9000)* Confirmation by examination and through provision of objective evidence that the requirements for a specific intended use or application have been fulfilled.

Value-based quality [ASTQB](*Ref: After Garvin)(See Also: manufacturing-based quality, product-based quality, transcendent-based quality, userbased quality)* A view of quality wherein quality is defined by price. A quality product or service is one that provides desired performance at an acceptable cost. Quality is determined by means of a decision process with stakeholders on trade-offs between time, effort and cost aspects.

Variable [ASTQB] An element of storage in a computer that is accessible by a software program by referring to it by a name.

Verification [ASTQB](*Ref: ISO 9000)* Confirmation by examination and through provision of objective evidence that specified requirements have been fulfilled.

Vertical traceability [ASTQB] The tracing of requirements through the layers of development documentation to components.

Volume testing [ASTQB]*(See Also: resource-utilization testing)* Testing where the system is subjected to large volumes of data.

Walkthrough [ASTQB]*(Ref: Freedman and Weinberg, IEEE 1028)(See Also: peer review Synonyms: structured walkthrough)* A step-by-step presentation by the author of a document in order to gather information and to establish a common understanding of its content.

White-box test design technique [ASTQB]*(Synonyms: structural test design technique, structure-based test design technique, structure-based technique, whitebox technique)* Procedure to derive and/or select test cases based on an analysis of the internal structure of a component or system.

White-box testing [ASTQB]*(Synonyms: clear-box testing, code-based testing, glass-box testing, logic-coverage testing, logic-driven testing, structural testing, structure-based testing)* Testing based on an analysis of the internal structure of the component or system.

Wideband Delphi [ASTQB] An expert-based test estimation technique that aims at making an accurate estimation using the collective wisdom of the team members.

ABOUT THE AUTHOR

After earning his B.A. in History, Paul Felten enlisted in the US Army, as an infantryman. While enlisted he earned the coveted Ranger tab and served two deployments as part of the 82nd Airborne. After his service obligation was complete he was offered an apprenticeship by Brian Pate, an expert medical device consultant helping medical device companies achieve compliance with all standards and regulations. Paul started manual testing, then began expanding to surrounding activities such as creating requirements, creating test cases, and creating test plans. Paul then taught himself test automation with Ranorex and through Ranorex, taught himself how to program in C#. He has helped clients achieve test compliance and efficiency for embedded software, desktop applications, mobile applications, cloud-hosted virtualization software, and more. Paul continues help clients with their verification needs, both manually and through test automation as Senior Validation Specialist for SoftwareCPR® and also teaches Ranorex training courses through ASPE, Inc.